CONVERSION OF MILITARY R&D

Conversion of Military R&D

Edited by

Judith Reppy

Associate Editors

Joseph Rotblat

John Holdren

Vsevolod Avduyevsky

First published in Great Britain 1998 by
MACMILLAN PRESS LTD
Houndmills, Basingstoke, Hampshire RG21 6XS and London
Companies and representatives throughout the world

A catalogue record for this book is available from the British Library.

ISBN 0-333-73871-3

First published in the United States of America 1998 by
ST. MARTIN'S PRESS, INC.,
Scholarly and Reference Division,
175 Fifth Avenue, New York, N.Y. 10010

ISBN 0-312-21802-8

Library of Congress Cataloging-in-Publication Data
Conversion of military R&D / edited by Judith Reppy ; associate editors, Joseph Rotblat, John Holdren, Vsevolod Avduyevsky.
p. cm.
Includes bibliographical references and index.
ISBN 0-312-21802-8
1. Military research. 2. Economic conversion. 3. Nuclear weapons industry. I. Reppy, Judith, 1937– .
U390.C66 1998
355'.07—dc21 98-8308
CIP

This book is printed on paper suitable for recycling and made from fully managed and sustained forest sources.

10 9 8 7 6 5 4 3 2 1
07 06 05 04 03 02 01 00 99 98

Printed and bound in Great Britain by
Antony Rowe Ltd, Chippenham, Wiltshire

Contents

Preface

The end of the Cold War brought the nuclear arms race to a halt and started the process of mutual reductions in military arsenals. Bilateral negotiations between the United States and Russia have culminated in the START agreements which, when implemented, will result in greatly reduced nuclear arsenals in these countries.

The Pugwash Conference on Science and World Affairs - where the nuclear issue was from the beginning on the top of the agenda - welcomed the START agreements, but considered them only as a first step towards the total elimination of nuclear weapons. A comprehensive study of the nuclear issue carried out by Pugwash resulted in the publication of a monograph, *A Nuclear-Weapon-Free World: Desirable? Feasible?*, which was translated into seven languages, and became an important contribution to the academic debate on the subject, as well as providing an input to the mass movements for a nuclear-weapon-free world. The efforts of Pugwash towards this goal received world recognition by the award of the Nobel Peace Prize in 1995.

While working towards the goal of the elimination of nuclear weapons, Pugwash has also been concerned about the actual process of disarmament, not only of nuclear weapons, but also of chemical, biological and conventional weapons. The ultimate goal of Pugwash is to eliminate the causes of disputes between nations and the tendencies to resort to force in settling them. This requires wide-ranging constraints on military R&D, since the identification of new weapon possibilities invariably generates irresistible pressures for their production and employment. For this reason, and also because of potential benefits in the civilian sector, there is an urgent need to transfer much of the money, skills, and production capacity from military R&D to peaceful R&D.

In the process of conversion it is important to ensure that the transition from military to peaceful applications is genuine and irreversible, i.e., that the change is not from one type of military R&D to another military use disguised as a peaceful project. It is also important to ensure that the staff of the military research establishment do not suffer undue hardships in the process, and that the skills and experience of the scientists are utilized in the proposed projects. To a certain extent this could be achieved by starting projects on the process of disarmament itself. 'The old poacher makes the best keeper.' The weapon-makers are often the most qualified to develop procedures for the efficient and safe dismantling and disposal of weapons, and to work out a robust verification system for arms control treaties.

These desiderata for conversion have not been met in practice: indeed, some untoward effects of disarmament measures have been noted. The initial expectation of 'peace dividends' from the reduction of military spending has proved illusory to a large extent. In some cases the funds saved have been diverted to pseudo-peaceful applications; in other cases they have been withdrawn completely, leaving many scientists without jobs, or without minimum salaries, causing real deprivation to them and their families.

In these respects, there is a radical difference between the situation in Russia and that in the United States. In the United States a gradual conversion away from weapons work has been going on for some time in the Los Alamos and Livermore National Laboratories, and these new research commitments have cushioned to a large extent the impact of the end of the Cold War. The new projects include arms control studies such as dismantlement of weapons and development of verification technologies. However, a much larger effort, and huge financial resources, have gone into projects which are presented as largely peaceful, but have a significant military component. An example is the National Ignition Facility. There is a strong opinion that its principal purpose is to retain a strong and active cadre of nuclear weapons experts, to be available should the need arise to restart a nuclear weapons program at short notice.

In Russia, by contrast, there has been a sudden, draconian cut in research funding in the former secret 'nuclear cities.' The work in the Chelyabinsks, Arzamases, etc. has come virtually to a standstill, and salaries of staff, in real terms, have become derisory. The international scientific community made an effort to provide some alleviation (though not necessarily in the interest of science in Russia) by helping a number of scientists to find jobs in other countries. Several governmental, as well as private, institutions, such as the Soros Foundation, provided for a time supplementary salaries for a significant proportion of the scientific staff engaged in non-military work. But for the majority of the personnel the situation remains tragic. This has led to fears that, in desperation, some scientists may resort to illegal steps with dangerous consequences for world security, such as stealing and selling abroad nuclear-weapons materials, or offering their intimate knowledge of the design of nuclear weapons to countries bent on the acquisition of such weapons. Rumors abound about such illicit transactions, although they have never been substantiated on a significant scale.

Other scientists have sought to put life back into the establishments by promoting new legitimate, yet undesirable, projects. An example is the 'peaceful' use of nuclear explosions. Such projects were topical several decades ago, but were abandoned when closer analysis showed that the projects could be carried out more safely or cheaply by the use of conventional explosives. Another, more recent, project is the use of nuclear explosives for the prevention of catastrophic collisions with heavenly bodies, such as asteroids and comets, by shooting large numbers of hydrogen bombs into, or near, them in order to deflect them from the Earth. The continuous maintenance of huge nuclear arsenals for an event of a very low probability is, however, clearly unacceptable in view of the much greater probability of these arsenals being used, inadvertently or deliberately, in a collision between nations. Any scheme which involves the continuous existence of weapons of mass destruction, even under the guise of peaceful applications, is against the Pugwash spirit, based on the social responsibility of scientists. Pugwash must seek solutions that are compatible with the social conscience of scientists.

It is largely due to such considerations that Pugwash embarked on the project 'Conversion of Military R&D to Peaceful Applications,' with the specific aim to seek ways to redirect the huge resources - economic, financial, industrial, scientific - now spent in military establishments, towards application in civilian life. The Pugwash project entailed holding several meetings in which scientists and scholars from a number of countries debated the ways and means of achieving a safe and beneficial transition to a world less heavily armed and more heavily committed to the alleviation of misery, hunger and disease in the world.

The outcome of these debates is this monograph consisting of well-documented chapters on specific aspects of conversion; the authors, scholars in the relevant areas, review the current situation, and suggest acceptable solutions.

The project was coordinated by John Holdren, Vsevolod Avduyevsky, and myself. The authors of the chapters come from the countries most directly concerned. Four of these - France, Russia, the United Kingdom and the United States - are nuclear weapon states. Germany is included because of the extensive study of the conversion problem carried out there. Apart from the authors of the chapters, the following took an active part in the debates: Professor Jack Boag, Professor Mike Cooley, and Dr. Adrian Mears, of the UK; and Professor Richard Garwin and Professor Catherine Kelleher of the United States. To all of them we wish to express our gratitude for the valuable contributions they have made.

We are most grateful to Judith Reppy for undertaking the arduous task of editing the papers and preparing the material in a form ready for publication. Professor Reppy also wrote the Executive Summary, a masterful digest of the individual ideas. She was ably assisted by Sandra

Kisner, whom we thank for her assiduous work on the manuscripts. Our thanks are also due to Tom Milne for preparing the final text for printing.

It is our hope that by publishing this monograph, Pugwash is making a useful contribution to the understanding of a problem of vital importance to peace and security in the world, as well as proposing practical steps towards its solution.

Joseph Rotblat

Notes on the Contributors

Jürgen Altmann is the project leader of the Bochum Verification Project at the Ruhr Universität at Bochum/North Rhine Westphalia, which does experiments on sensor detection of land and air vehicles. Since 1986 he has worked on scientific aspects of arms control and disarmament. He has published major studies on laser weapons, ballistic missile defense, and verification.

Yuri Andreev is Principal Research Officer at the Peace Research Centre, Institute of World Economy and International Relations, Russian Academy of Sciences.

Nicolai A. Anfimov, a corresponding member of the Russian Academy of Sciences and of the International Academy of Aeronautics, is a well known leader in space engineering, especially in studies of aerogasdynamics, thermal processes, and ground testing of rockets and spacecraft.

Vsevolod S. Avduyevsky is a leading scientist in the field of rocket and space technology. He has more than 40 years of experience in development and systems analysis of rocket and space systems. He is President of the Russian Society for Promoting Conversion.

Evgeny N. Avrorin is Scientific Head of the Institute of Technical Physics, Snezhinsk (Chelyabinsk-70). He has published a series of papers on the behavior of various substances in extreme environments and on laser thermal nuclear fusion, as well as on the subject of nuclear disarmament and world stability.

Philip Gummett is professor of Government and Technology Policy at the University of Manchester, UK, and is academic coordinator of a European-wide network (CREDIT) that researches policy issues concerning defense technology.

Jean-Paul Hébert, with a doctorate in economic theory, is an economist specializing in defense industries and arms sales. He is a member of the Groupe de Sociologie de la Défense (GSD) in the École des Hautes Études en Sciences Sociales (EHESS) in Paris. He has recently published *Production d'armement: la mutation du système française* (la documentation française, 1995).

John Holdren is the Teresa and John Heinz Professor of Environmental Policy and Director of the Program on Science, Technology, and Public Policy at the John F. Kennedy School of Government, Harvard University. He is also Chair of the Committee on International Security and Arms Control of the U.S. National Academy of Sciences, and a member of President Clinton's Committee of Advisors on Science and Technology.

Ray E. Kidder was a physicist at the Lawrence Livermore National Laboratory from 1956 to 1990 working on nuclear weapons concepts and design, laser-driven fusion and isotope enrichment, and nuclear weapons reliability, safety, and arms control. He is a fellow of the American Physical Society and winner of its Leo Szilard award.

Wolfgang Liebert is a physicist and research fellow at the Interdisciplinary Research Group on Science and Security Policy Group (IANUS) at the Technical University at Darmstadt. His main field of research is nuclear non-proliferation, qualitative arms control, and technology assessment. He is a member of the International Network of Engineers and Scientists Against Proliferation (INESAP).

John Tepper Marlin is former director of the Conversion Information Center at the Council on Economic Priorities (CEP) and is a member of the Executive Committee of Economists Allied for Arms Reduction. He co-authored with Betty G. Lall *Building a Peace Economy: Opportunities and Problems of Post-Cold War Defense Cuts* (Westview Press, 1992). Since January 1994 he has served as Senior Policy Adviser to the New York City Comptroller.

Ivan Maslennikov is a professor at the Military Dzerjinsky Academy, Moscow, and an associate of the Peace Research Institute Moscow (PRIM), Russian Academy of Sciences.

Tom Milne is a researcher at the London Office of the Pugwash Conferences on Science and World Affairs. His research is on UK nuclear weapons policy.

Götz Neuneck is a physicist and research fellow at the Institute of Peace Research and Security Policy at the University of Hamburg. His research topics are non-proliferation, missile defense, preventive arms control, high technology, and strategy. He is a member of the International Network of Engineers and Scientists Against Proliferation

(INESAP) study group, and the Center of Natural Science and International Security (CENSIS) in Hamburg.

Mikhail Ilyich Nikitin is Director of the Central Research Institute for Economics, Informatics and Control Systems, specializing in economic and industry management systems. He is Professor and member of the International Informatization Academy.

Judith Reppy is an economist specializing in defense technology issues. She is a member of the Department of Science and Technology Studies at Cornell University and has served as Director of Cornell's Peace Studies Program.

Joseph Rotblat is emeritus Professor of Physics at the University of London, and co-recipient of the 1995 Nobel Peace Prize. Previously he has served as Secretary-General and President of the Pugwash Conferences on Science and World Affairs.

Jack Ruina is emeritus Professor of Electrical Engineering at the Massachusetts Institute of Technology. He has formerly served as Director of the Advanced Research Projects Agency, U.S. Department of Defense; as President of the Institute for Defense Analysis; and as a Senior Consultant to the White House Office of Science and Technology.

Jürgen Scheffran is a physicist and research fellow at the Interdisciplinary Research Group on Science and Security Policy Group (IANUS) at the Technical University of Darmstadt. His main fields of research are mathematical modelling and stability of complex systems, missile proliferation, missile defense, and arms control. He is a member of International Network of Engineers and Scientists Against Proliferation (INESAP).

Nicolai A. Severtsev, vice admiral, was involved in the development of spacecraft and rockets. Now he is specializing in research on system safety.

Peter Southwood is a writer and consultant who obtained his doctorate on arms conversion and the UK defense industry from the Department of Peace Studies, University of Bradford. He was formerly a member of the Steering Committee of the Parliamentary Office of Science and Technology study on defense conversion.

Wim A. Smit is associate professor and director of the Centre for Studies of Science, Technology and Society at the University of Twente, the Netherlands. He has published on nuclear weapon proliferation, processes of military technological development, and technology assessment.

Josephine Anne Stein is an engineer, formerly with the US Office of Technology Assessment, now senior research fellow with the Programme of Policy Research in Engineering, Science and Technology (PREST), University of Manchester, UK, and manager of PREST's London office.

Adam Yarmolinsky served in the Pentagon and the US Arms Control and Disarmament Agency in the Kennedy, Johnson and Carter administrations, and has written extensively on international security issues. He is Regents Professor of Public Policy in the University of Maryland system.

Conversion of Military R&D

Executive Summary

INTRODUCTION

Throughout the Cold War the United States and the Soviet Union, along with a number of other industrialized countries, expended large resources for military research and development (R&D). National security was invoked to justify the expenditures, which yielded ever more technologically complex and lethal weapons on both sides. But now the Cold War is over, and the security landscape in Europe has been transformed by the dissolution of the Warsaw Treaty Organization and the breakup of the Soviet Union. There is an opportunity in all countries to reappraise the benefits and costs of military R&D, to consider the extent to which the laboratories and other institutions engaged in military R&D are still needed, and to plan for conversion to civilian uses of the resources that are being released by the decreases in military spending.

Conversion is here defined to mean the shift of personnel, equipment or facilities previously utilized for military R&D to R&D directed toward civilian aims. Other related concepts are the transformation of military R&D to new tasks, e.g., nuclear stockpile stewardship; diversification by adding civilian-oriented projects without reducing military R&D; reprioritization of R&D to civilian activities without employing the resources previously used in military R&D; and downsizing, again without employing the resources made idle by cuts in military R&D. The prospects for true conversion vary among countries and across industries, and realistic planning for conversion requires that individual circumstances be taken into account. This book addresses the issues involved in conversion of military R&D with contributions from scientists and analysts from Russia, the United States, France, Germany and the United Kingdom.

PART I: SETTING THE CONTEXT

Spending for military R&D must be seen in the context of other spending for research and development and of total military spending. Available data summarized in Chapter 1 show that in the early 1990s a few countries - the United States, the United Kingdom, France and Sweden - had disproportionately large expenditures for military R&D, ranging as high as 59 per cent of government R&D and 14 per cent of military spending in the United States. In effect these countries have protected their military R&D programs, even as total military budgets have fallen. By contrast, spending in Russia, which inherited most of the Soviet Union's military R&D infrastructure, has dropped sharply as a result of both policy decisions and the chaotic fiscal situation after 1990. Although the data for spending on military R&D must be viewed with caution because of problems with definitions and coverage, they do provide an indicator of the opportunity costs of military R&D and the scope of the conversion task in individual countries. Additional information on R&D spending is provided in the individual country studies in Part II.

The *benefits* from military R&D can be summarized under the heading of military benefits - e.g., technological advances that translate into more effective weapons plus the general sense that an investment in new technology maintains the technology base and provides a hedge against future security threats - and civilian benefits - e.g., a general stimulus to technology from military spending and the possibility that specific military developments will have significant civilian applications. In recent years, with an increase in the importance of civilian-based technologies, especially information technologies, the 'spin-off' model of technology transfer from the military to civilian sectors has been replaced by a focus on 'dual-use' technologies. As a consequence the boundary between military and civilian technology has become more fuzzy, and the direction of technology flow more difficult to determine. The *costs* of military R&D include, on the military side, the budgetary costs of increasingly sophisticated weapons and the danger of technological arms races and, on the civilian side, the opportunity costs of allocating skilled workers and facilities to military rather than civilian purposes, especially the possibility that countries that invest heavily in military R&D will lose out to competitors in civilian markets.

Chapter 2 reviews the arguments and evidence for a positive or negative impact of military R&D spending on the civilian economy and finds the evidence inconclusive, at least for the United States and Britain, the most-studied countries. Owing to the inherent difficulties in identifying and measuring the costs and benefits of R&D and the complexity of the relationship between the military and civilian sectors, these studies yield only ambiguous results. For example, although defense spending drains technical personnel from the civilian sector, it also stimulates the supply of scientists and engineers through subsidized scholarships and the promise of lucrative employment. The chapter concludes that the economic arguments for and against military R&D are not decisive; the security calculus should prevail. Current conditions provide an opportunity for reducing military R&D and the rate of military technological change, but the increased importance of dual-use technologies may transform the security issues as well as the economic ones.

PART II: CONVERSION OF MILITARY R&D

The chapters in Part II provide an overview of the issues involved in converting military R&D to civilian purposes and more detailed studies of the problems present in specific countries. Conversion implies a radical transformation of the war economy and military institutions to peaceful purposes; conversion of military R&D, therefore, must be viewed in the context of sweeping changes in both military activities and civilian society. In the short run conversion is part of changes in national defense policy and procurement planning at the end of the Cold War. A longer term perspective, however, points to disarmament: successful conversion is essential to lessen opposition to disarmament stemming from fear of job losses and from security concerns. A variety of factors that affect the prospects for conversion are identified in Chapter 3. They occur at the macro, micro and regional levels, which underscores the importance of placing conversion firmly in the context of an overall policy framework at the national level. A fourth dimension is the possibility of international cooperation, especially to assist the countries of Eastern Europe and the successor states to the Soviet Union. Such cooperation should be developed collaboratively and based on an understanding of the differing economic and social contexts in these countries.

Following these general guidelines, Chapters 4 and 5 discuss the context for conversion in the United States and Russia, and Chapters 6 and 7 describe the conversion policies of the two countries. Chapters 8–10 describe the political and economic context for conversion in France, Germany and the United Kingdom.

In the U.S. case, the problem lies in the different cultures of the military and civilian sectors, as described in Chapters 4 and 6. Under the Clinton administration there has been a greater willingness to consider conversion as part of a larger project of government investment in technology development than under the previous Republican administrations. The programs proposed, however, have focused on dual-use technologies and been lodged in the Department of Defense, and critics argue that they are simply a means of continuing support for military-related technologies under a different rubric. Even these programs, limited as they are, have been attacked by the Republican Congress elected in 1994. Other conversion efforts by state and local governments are more promising, and there are examples of defense firms engaging in successful conversion.

In Russia conversion must be seen in the context of a transformation of the entire economic system to a market economy. Prospects for conversion depend on striking a balance between continuing security needs, and the need to reduce the over-expanded military sector through diversification, without sacrificing the technological excellence of the science-intensive industries through inappropriate civilian projects. Conversion in Russia has moved through several stages, which are summarized in Chapter 7. The plan put forward under perestroika in 1987 was based on a centralized state program; under this plan specialized military R&D resources were supposed to aid in raising the technical level in specific civilian industrial projects, which were brought into the defense sector. Conditions changed in 1989 when steps towards a market economy were introduced. Military science enterprises, however, remained part of the state sector, although with sharply reduced budgets and without any clear policy direction. Since 1992 there has been a more settled situation in which state enterprises have been encouraged to develop business plans for conversion and to locate investment sources. As establishments have diversified into civilian work, however, the state has only partially compensated for the fall in military orders, and there has been considerable brain drain of qualified staff. There is no shortage of imaginative ideas of ways to utilize the military technology base, but there is insufficient finance and other infrastructure to implement the conversion projects. Some enterprises have succeeded in shifting their scientific resources to new civilian applications, however, and there is scope for considerable progress if international collaboration can be increased.

The situation in France is particularly interesting. France has invested heavily in military R&D, which lies at the heart of French strategic priorities, and R&D's share in the military budget has grown since the mid-eighties. France desires to maintain a high level of military technical vigilance and an autonomous capability in military R&D in order to protect its sovereignty. Moreover, it sees technological competence - even if only potential rather than realized - as necessary in order to maintain France's standing in international negotiations. R&D conversion will therefore be difficult despite the economic constraints that the industry faces; there is no official conversion plan. Nevertheless, a creeping conversion is taking place in the arms industry as employment and production fall in response to declining domestic orders and foreign sales. The process is uneven across sectors, with the nuclear sector relatively protected so far, while considerable diversification has occurred in the aeronautics industry. The need for conversion is most pressing in the area of ground weapons. The importance of military R&D to national strategy, however, suggests that there will be little progress towards conversion, except perhaps in the nuclear sector in response to changes in nuclear doctrine.

Germany, by contrast, has in the post-World War II era emphasized integration into European and NATO security structures rather than autonomous capabilities. With the end of the Cold War and the disappearance of the traditional security threat, there is an increased interest in out-of-area operations and peace-keeping, and German military forces have adopted new programs to increase mobility and relocation abilities. Overall, the size of the military and its associated industry has fallen dramatically since re-unification. To retain some military technology capability, Germany has turned to international collaboration, especially within Europe, and to a policy of integrating civil technology into military developments. This policy leads to a strategy of diversification in R&D activities, rather than irreversible conversion of scientific and technological resources, and poses the danger of a militarization of R&D in Germany that would undermine efforts for conversion and disarmament in the long run.

The United Kingdom, like the other major performers of military R&D, is experiencing cutbacks from its previous high levels of spending in response to the end of the Cold War. Nuclear weapons remain at the center of UK strategy, and the Atomic Weapons Establishment's laboratories are important sites for British science and technology. It is government policy to protect this technology base despite the lack of any new nuclear programs in the foreseeable futures; one possible area for diversification would be in developing technologies in support of nuclear arms control agreements. The non-nuclear research establishments have been amalgamated in the Defence Evaluation and Research Agency, which is now managed as a trading fund along commercial lines. UK policy emphasizes competitive contracting for military R&D and increased links to the civilian sector; there are no formal policies for conversion of military R&D. Defense industrial firms are being encouraged to diversify into civilian technologies and markets, but an industry consensus holds that there is little scope for conversion outside the field of aerospace, and there it is limited by excess capacity in existing commercial capabilities. The United Kingdom also has significant capability in the technology for chemical and biological weapons defense and, in this field at least, devotes considerable resources to supporting arms control measures, through, e.g., work on verification technology.

These case studies taken together suggest that, although the costs of military R&D are considerable, governments (other than Russia) with major military R&D programs are reluctant to pursue active policies for conversion. There is a general shift, instead, to promotion of dual-use technologies as a way of maintaining the defense technology base while encouraging defense firms and laboratories to be less dependent on defense spending. The implication of this policy on long term conversion and disarmament is, however, problematic, since it may result in a militarized civilian R&D base.

PART III: SPECIFIC OPPORTUNITIES AND CHALLENGES

The national nuclear weapons laboratories in the United States and Russia present special opportunities and problems for conversion because their histories and missions are so deeply identified with the nuclear rivalry that was the most threatening aspect of the Cold War. By contrast, the electronics and space industries have outgrown their military origins and enjoy substantial civilian markets; nevertheless, their remaining military sectors face considerable obstacles to conversion. The chapters in Part III address the specific opportunities and challenges that characterize these laboratories and industries.

The national nuclear weapons laboratories of the United States - Los Alamos National Laboratory and Lawrence Livermore National Laboratory - have lost a substantial part of their mission with the end of the Cold War. There are currently no new nuclear weapons under development or planned for the future, and a Comprehensive Test Ban Treaty has been negotiated and opened for signature. Nevertheless, the laboratories have argued for maintaining and modernizing their test facilities as a hedge against an uncertain future. 'Stewardship' of the shrinking stockpile has become the main weapons-related mission of the laboratories, along with technical support of arms control agreements and nonproliferation policy and nuclear waste management. These missions do not justify the budgets and levels of employment that the labs have enjoyed in the past, and there is considerable uncertainty as to their future. The laboratories face the challenge of finding alternative missions that will both provide employment for their rather specialized scientific and technological resources and also command national support. The technology transfer programs that they embraced enthusiastically in the early 1990s are not enough, and have been criticized as by some as inappropriate. In short, while there is near consensus on a the need for a small, but robust national capability for the nuclear weapons stewardship function, there is no agreement on other long-term missions for the U.S. laboratories. The laboratories' strategy appears to be to emphasize the continuing importance of their defense mission in the hope of protecting their budgets from deeper cuts.

In Russia there are also two main nuclear weapons laboratories, Arzamas-16 and Chelyabinsk-70, which have co-operated and competed with each other in the development of nuclear weapons. Discussion of conversion possibilities over the years had led to an understanding that no single civilian task would equal the nuclear weapons mission's requirement for scientific capacity and complexity. Instead, a range of opportunities have been identified in areas such as nuclear power engineering, environmental contamination, medical devices and communications devices. There are numerous barriers to be overcome, however, including reversing the culture of secrecy to encourage contacts with other scientists at home and abroad; reorganizing project teams to increase flexibility; restructuring the laboratories around projects rather than disciplines; and making the legal changes necessary for the laboratories to cooperate with other state and private enterprises. The greatest challenge has been to secure sufficient financing for research projects, which necessarily do not promise a quick payoff. The nuclear centers have fared better than many Russian research organizations in state funding and in international aid, which now stands at approximately 10 per cent of the centers' budgets and helps to integrate their scientists into the world scientific community. There is a danger, however, that these international contacts will be used exploit Russia as a source of low cost technology and to create dependencies on Western advanced components. In the long run conversion will best serve Russia's needs if it is based on indigenous development of the scientific potential and production facilities for civilian goods, with relations with the West conducted on a basis of equality.

Conversion should be easier in industries like electronics and space, where there are sizeable civilian markets, than in the specialized nuclear weapons laboratories. Closer examination of the U.S. military electronics industry, however, reveals that it shares many of the characteristics of other parts of the military industrial complex, including an emphasis on high performance in demanding environments and a tolerance for high development costs that would be unacceptable in a commercial business. It is likely that government initiatives to encourage new technology will benefit the commercial side of the industry, but not result in conversion of the military sector.

The Soviet space industry had reached a high level of technical development during the Cold War; now it is trying to apply its capabilities to the civilian sector. There is great potential demand, especially for satellite communications systems and remote-sensing satellites for assessing natural resources and monitoring the environment. Technology could be spun off to address other civilian technology needs in industries such as mining, atomic energy and transportation. Testing facilities developed in the space complex - for example, wind tunnels - can also benefit civilian research. Against this potential for conversion, one must place the current disarray in the rocket and space industry, which has experienced several years of reduced funding from the government and is losing some of its scientists to more lucrative jobs outside the space sector. Although this is a kind of conversion, it comes at the cost of a lower creative capability; similarly, the production of low technology goods in the unused production facilities of the industry has reduced its overall technical capacity.

These cases taken together suggest an interesting paradox. In Russia conversion is an absolute necessity, but it must take place in the context of large scale economic and political restructuring and thus faces many obstacles. There is an impressive array of creative ideas for conversion of the defense research establishment, but success so far has been illusive. In the United States, by contrast, the obstacles are fewer and less forbidding, but the sense of urgency is also less, so that conversion in the sense of redeployment of military resources to civilian projects is also rare. Instead, in both countries the economic costs of the Cold War continue to be felt, as specialized facilities lie idle and trained scientists and engineers are underemployed.

PART IV: THE WAY FORWARD

Military R&D, although justified in the name of national security, more often creates insecurity through its role in arms races. Conversion of military R&D thus is desirable on security as well as economic grounds; indeed it is a necessary component of a future international regime of disarmament and confidence building. The chapters in Part IV discuss possible paths to such a regime and analyze the specific controls on R&D that would contribute to comprehensive security without unduly limiting desirable technical progress.

Arms control agreements in the past have emphasized quantitative ceilings rather than qualitative controls that would limit military R&D. Qualitative controls would provide a kind of preventive arms control that would avoid destabilizing innovations and address the problems of proliferation of high technology weapons. There is a tension between the goals of conversion through dual-use programs, which are most suited to the early stages of R&D when the application is not yet determined, and preventive arms control, which must deal with that very indeterminacy and somehow prevent the continuation of military projects under a civilian guise. These goals are consistent only in the context of a comprehensive disarmament process and shift to non-violent methods of conflict resolution. Such a regime would favor transparency in R&D programs and reduce the demand for new weapons. First steps towards such a regime include a comprehensive nuclear test ban and extension of safeguards to all nuclear facilities; more extensive controls on missile technology and other means of delivery; and extension of arms control measures to the area of conventional weapons. The challenge is to identify markers that can distinguish military from civilian-oriented technologies, so that only dangerous research and testing is prohibited. Limits on R&D could be supplemented by a variety of confidence-building and verification measures that could be introduced progressively as the disarmament regime became established. Such a regime would be enhanced by developing an international scientific capability to oversee disarmament and promote conversion of military R&D on a cooperative basis.

Advocates of a preventive arms control regime must address the central issue of distinguishing military from civilian R&D. Chapter 16 presents a series of case studies of dual-use technologies and their institutional settings. It argues from these examples that research and development can be 'steered' through careful analysis of institutional settings and intentions; that is, the social context can be monitored on a case-by-case basis for clues as to the probable application of the technology under development, and appropriate trade-offs negotiated between civilian benefits and threats to international security. Ideally such monitoring should be done by an international body and applied equally to all countries to avoid the charge of discrimination that has plagued the nuclear non-proliferation regime.

Introduction

John Holdren and Judith Reppy

The sudden end of the Cold War in 1989–91 brought widespread expectations of a 'peace dividend' to be reaped from the redirection of economic resources that until then had been devoted to the massive military establishments, in both East and West, spawned by four decades of confrontation. The peace dividend was expected to arise in two ways: reductions in military spending overall (which would allow governments to increase spending on social programs, or to reduce deficits and pay down debt, or to decrease taxes and thereby stimulate increased economic activity in the private sector); and from the 'conversion' of resources with high economic leverage to the pursuit of increased quantity, quality, and variety of civilian goods and services and hence improved quality of life. These resources included specifically the productive capacities of high-technology defense-oriented firms and the inventiveness of defense-oriented research-and-development (R&D) establishments.

Nearly a decade after the dismantling of the Berlin Wall, however, the barriers to the realization of the peace dividend have proven to be considerably more difficult to overcome than most people had supposed. First of all, military budgets have not fallen as rapidly, in some cases, as was expected, partly because of the high costs of relocating forces and dismantling of surplus weapons, partly because of a tendency to 'hedge' against the re-emergence of tensions on the old Cold War axes by not reducing forces as sharply as might have been expected, and partly because of a tendency to reorient forces – rather than downsizing them – toward real and potential regional conflicts that the ending of the Cold War did nothing to stop (or even, in some cases, encouraged). Second, the specifics of the implementation of the conversion to civilian ends of defense production and defense R&D have proven to be beset with problems and obstacles in the technical, economic, and political domains, including but by no means limited to the same 'hedging' and reorientation inclinations just mentioned in connection with overall military spending.

Our focus in this book is on the research-and-development dimension of this conversion issue: on the motivations, opportunities, obstacles, prospects, and problems associated with converting military R&D

resources and activities to the pursuit of civilian ends. Military R&D activities are important to the military as the source of future weapons and weapons-related capabilities; they are important to civilian society because they absorb the talents of large numbers of scientists and engineers – talents that in principle could be applied to civilian needs. Conversion of military R&D thus presents a double opportunity: the opportunity to moderate the sorts of technological arms race that characterized the Cold War, and the opportunity to enhance the civilian technology base to the direct benefit of human well-being.

Conversion of military R&D is not, however, an easy task, as the chapters in this book make clear. There is no panacea or general prescription for conversion. Instead, one must be prepared to do the detailed intellectual work needed to understand the situation in each country – and in all the relevant sectors of military and civilian R&D activity – and to fashion practical conversion plans accordingly; and one must be prepared to do the challenging political work required to build support for carrying out such plans. This book is intended to be a contribution to that effort.

In the rest of this introduction, we summarize some of the major themes and key ideas that are developed in greater detail in the sixteen chapters to follow. Our treatment is organized under six headings: the context and motivation for conversion of military R&D; definitions and variations of conversion and related concepts; assessing conversion opportunities; barriers and policies for overcoming them; dual-use issues; conversion issues in specific countries and sectors; and concluding observations.

CONTEXT AND MOTIVATION FOR CONVERSION

For reasons elaborated in Chapter 1 – including secrecy, ambiguous definitions, exchange-rate distortions, and more – it is not possible to obtain very accurate estimates of worldwide spending on military R&D (or, for that matter, on military activities overall). In round numbers, however, it is not far off to say that world military spending at the end of the Cold War was equivalent to about US$1,000 billion per year – that is, about 5 per cent of a world economic product around US$20,000 billion per year – and that expenditures on military R&D probably amounted to something in the range of 10 per cent of total military spending, hence around US$100 billion per year.[1] The industrialized countries that generate about 85 per cent of the world's economic product and do most of the world's research and development spend typically 2–3 per cent of their gross domestic products (GDPs) on all forms of R&D combined (including private-sector R&D expenditures),[2] which suggests that total world R&D expenditures in the late 1980s were perhaps US$400–500 billion; thus military R&D would have accounted for 20–25 per cent of the total.[3]

The opportunity cost of military R&D, from the viewpoint of the society as a whole, is probably higher than is suggested by its aggregate share of all R&D spending. This is so because the 20–25 per cent of total R&D that is government-funded military R&D comes from the pool of funds – government as opposed to private-sector R&D funding – that constitutes most of the money available for those types of R&D in society's interest that the private sector has little incentive to support. These include: basic research; applied research and development on nondefense 'public goods' such as environmental quality, health, education, information and transportation infrastructures, disaster avoidance and mitigation, and crime prevention; and applied research and development on technologies, such as fusion energy, with such long development times that the prospect of an economic return on the research investment is too remote to attract substantial private-sector involvement. To the extent that high levels of spending on military R&D have competed with these types of civilian R&D investment, they have cost society dearly.

The situation in the United States, which at the end of the Cold War accounted for 30–35 per cent of the world's total R&D spending and perhaps 40 per cent of military R&D spending, provides an illustration of this problem. In 1990, U.S. government expenditures on military R&D were $40 billion, 62 per cent of all government R&D. By way of comparison, the government's 1990 R&D expenditures for health were only $8.3 billion, for energy supply and energy efficiency only $1.9 billion, for natural resources and environment only $1.8 billion, for transportation only $0.9 billion, for agriculture only $0.8 billion. It is particularly striking that the U.S. government's expenditures on energy, natural resources, and environment R&D totalled less than a tenth of its expenditures on military R&D, inasmuch as the threats to U.S. security in the next century could probably be more effectively abated by building the energy and environmental basis of sustainable global prosperity than by building better weapons.[4]

A major motivation for interest in conversion of military R&D, then, is the hope and expectation that R&D resources – money, people, facilities – could actually be transferred from military purposes that now seem less compelling than before to civilian purposes that seem more compelling than ever. Those civilian purposes include not only 'public goods' in the categories already mentioned, but also innovation in support of better consumer products and higher productivity in the private sector. In the latter connection, although considerable private investment in R&D is motivated by the expectation of short-term economic returns to the investors, there is scope for government R&D as well; public-sector R&D investments in general, as well as particular R&D skills and equipment located in national laboratories, government research institutes, and government-funded university research centers, can make important contributions to industrial innovation and competitiveness.[5]

Put in these terms, the current discussion about conversion of military R&D is just a particular way of talking about the fundamental and longstanding dilemma associated with allocation of the limited R&D resources at society's disposal. R&D is an indispensable ingredient of military capability, of industrial productivity and competitiveness, and of important aspects of human well-being including public health, environmental sustainability, and the advancement of basic understanding of ourselves and our universe. How, then, should R&D resources be allocated among these various aims?

While there is as yet no generally accepted prescription for making this allocation, there is certainly a rather widespread conviction that the share of R&D resources being devoted to military aims is too high in relation to the share being devoted to the other aims. That conviction has been strengthened by the end of the Cold War. It has been strengthened further by the understanding that 'spinoff' of military technology to nonmilitary aims is no longer as effective as it once was, and by the belief that the increasing globalization of the world economy has increased the importance to individual countries of not neglecting the development of their civilian technology base while focusing on the military side.

As discussed in this volume in the chapters by Gummett and Stein (ch. 2) and Smit (ch. 16), among others, the declining salience of spinoff from military R&D into other sectors has been the result of several factors.[6] In some cases, civilian requirements for technological performance have outstripped military ones (as appears to be the case with electronics), so that civilian R&D is spinning benefits onto the military sector rather than vice versa. In other cases, such as directed-energy weapons and high-performance military aircraft, the military's demand for extremely specialized capabilities without regard to cost has driven military R&D in directions less and less likely to be relevant to civilian applications; the salience of this factor is underlined by the high proportion of total defense R&D that falls in the category of development of specific weapons systems (about 87 per cent of U.S. defense R&D

spending in 1990, for example), as opposed to basic and applied research of wider potential relevance. In still other cases, the luster even of past cases of spinoff has been dimmed by arguments suggesting that technologies spun off from the military sector diverted civilian development from more promising pathways, or that the spun-off technologies proved to be dangerous hammers in search of nonexistent nails. (The first argument has been made, for example, about the spinoff of light-water reactors from military naval propulsion into civilian nuclear electricity generation, which might have been better served by reactors of different types; the second argument has been made about peaceful nuclear explosives.)

The globalization of the world economy means that trade barriers can no longer be relied upon to protect from foreign competition those domestic industries that use substandard production technologies and/or make substandard products. That means, in turn, that all countries are inescapably locked in a civilian technological competition in which those countries that spend less on military R&D in order to spend more on civilian R&D are likely, all else being equal, to gain the advantage. Of course, all else is not always equal, but there is enough plausibility in this basic argument to have caused great concern in countries, such as Russia and the United States, where public spending on R&D has been skewed toward the military side and whose major competitors and trading partners, notably Japan and Germany, have skewed their public R&D spending in the other direction.[7]

Additional motivations for the conversion of military R&D, besides the desire to increase the R&D resources available for nonmilitary aims, are often mentioned. These motivations, which are treated by a number of the authors of this book, include:

(a) avoiding unemployment, wastage of capabilities, social unrest, and temptations for weapons scientists and technologists to sell their skills in the world market (all of which would result from defense cutbacks if these were NOT accompanied by conversion);
(b) attracting to useful careers in science and technology a larger fraction of the brightest young people, some of whom currently take other directions because they do not wish to do military work and do not see nonmilitary R&D as adequately supported;
(c) contributing to a reduction in military expenditures overall in order to free these resources for other purposes besides nonmilitary R&D, such as consumer goods or debt reduction;

(d) reducing the role of military R&D in driving costly military procurements and destabilizing arms competitions; and
(e) building confidence that a degree of post-Cold-War demilitarization is both intended and not easily reversed.

It needs to be emphasized, however, that goals (a) and (b) are the only ones of these that strictly require conversion *per se*, which as elaborated below we take to mean redirecting specific R&D personnel, equipment, or facilities from military to civilian ends: the goal of money-saving (c) is explicitly *not* the same as conversion, while the goals of arms limitation (d) and confidence-building (e) do not require conversion in this strict sense but could be equally well served by mere *reductions* in military R&D without redirection of the military R&D resources to civilian ends. The link between goals (d) and (e) and actual conversion is rather more subtle, consisting in the proposition (as discussed by a number of the authors in this book) that political and economic forces might make reductions in military R&D unlikely - or, if achieved temporarily, then unlikely to be permanent - *unless* these occur by conversion.

It also needs to be emphasized that, all of the foregoing motivations notwithstanding, few people would argue in favor of the conversion of *all* military R&D. For one thing, the parts of military R&D that are basic research and applied research not directed to specific weapons systems are not necessarily less useful to the societal R&D enterprise than the same work would be if done under a civilian label. The frequent lack of fungibility of monetary resources across the military and civilian categories of government spending means that reduction of military spending in these R&D categories might well not be reflected in an increase in the corresponding civilian categories. As Gummett and Stein point out in chapter 2 of this volume, military R&D in most countries is considered part of the military budget, not part of the R&D budget, so that reductions in military R&D are more likely to imply increases in other categories of military spending than increases in other categories of R&D.

Second, as pointed out by Yarmolinsky in chapter 4, military R&D can be 'cost-shrinking' (by devising cheaper ways to perform existing missions) as opposed to 'cost-expanding' (by developing costly technologies to perform new missions). This means that military R&D expenditures, appropriately directed, can exert downward pressure on the much larger procurement and operations-and-maintenance categories of military spending,[8] potentially generating reductions in overall military spending greatly exceeding the cost of the R&D.

Third, it is at least arguable (and argued by a number of authors in this book) that military R&D provides a less expensive, less alarming, less provocative, and less destabilizing means of maintaining confidence in one's military security – including hedging against uncertainty about future security threats – than attempting to achieve similar ends by actually deploying weapons or stockpiling reserves of them. Of course, the validity of this argument depends on societies finding reliable ways to interfere with the usual progression of the 'technological imperative,' through which the well known characteristics of scientists (wishing to test their ideas), engineers (wishing to develop working prototypes), defense manufacturers (wishing to build a lot of what works and can be sold), and military leaders (wishing to buy whatever might convey superiority) tend to lead inexorably from research to development to deployment.

In any case, this diversity of rationales and complications relating to conversion of military R&D underlines the need for clarity about exactly what is included in this concept and what is not, so we turn now to these matters of definition.

DEFINITIONS AND VARIATIONS OF CONVERSION AND RELATED CONCEPTS

We think the term 'conversion,' as applied to military R&D, should be reserved for circumstances in which personnel, equipment, or facilities previously being utilized for military R&D are shifted to R&D directed toward civilian aims. This phenomenon is to be distinguished from the following related but different concepts:

- 'Transformation' of military R&D refers to alterations in the pattern of military R&D activities to respond to the transformed post-Cold-War security environment. This category of change includes increased R&D efforts related to: safe and secure transport, storage, dismantlement, and ultimate disposition of surplus weapons (especially nuclear and chemical weapons, which pose particularly demanding problems); maintenance, in the absence of nuclear-explosive tests, of the safety of the residual nuclear weapons in the stockpiles of states possessing these, for as long as such weapons are retained; technologies for detecting and countering proliferation of weapons of mass destruction to additional states and subnational groups; other verification and arms-control missions; and countering terrorism. A post-Cold-War transformation in military R&D might also entail shifts from cost-expanding to cost-reducing R&D, as mentioned above.

- 'Diversification' refers to adding civilian-oriented projects to the mix of R&D activities undertaken by a laboratory, institute, or firm by acquiring additional personnel, equipment, and facilities appropriate to these new activities, rather than by redirecting the application of personnel, equipment, or facilities formerly devoted to military R&D – that is, adding civilian activities without reducing military R&D activities. Of course, diversification may establish a foundation within a laboratory, institute, or firm from which actual conversion may subsequently take place in the form of internal shifts of personnel and equipment from its military to its civilian projects.
- 'Reprioritization' refers to shifting funds out of military R&D activities and into civilian R&D, without, however, using the same people, equipment, and facilities to do the additional civilian R&D as had been engaged in doing the military R&D that has been reduced.
- 'Downsizing' refers to a shifting of resources out of military R&D without corresponding increases in civilian R&D. Such R&D downsizing can take the form of resources being transferred from military R&D to non-R&D military activities, such as procurement or operations; or it can be part of a downsizing of the military sector as a whole, in which military R&D shrinks without corresponding increases in non-R&D military spending (or, indeed, with parallel decreases in the latter).

It is also important to distinguish, of course, between conversion of military R&D and conversion of military production activities. As noted above, production claims a much larger share of military spending than does R&D, and hence represents, in principle, a larger pool of economic resources that could potentially be transferred into the civilian economy; we believe, however, that the types of resources at stake in the conversion of military R&D, beyond mere money, have larger potential leverage on the state of the world than is the case with conversion of military production, and although the latter form of conversion raises some interesting issues we do not deal with them in detail in this book.

Among the categories of transforming, diversifying, reprioritizing, downsizing, and converting military R&D, transformation is clearly the easiest and least traumatic, and it has some significant arms-control and confidence-building benefits. But it is also the least relevant to wider (nonmilitary) social needs. Much of the change that has occurred since the end of the Cold War at the national nuclear-weapons laboratories of the United States and Russia (described here by Kidder, ch. 11, and

Avrorin, ch. 12) falls in this category. Maintaining nuclear-weapons capability is one of the goals of the transformed programs, so it is not surprising that they do not require a major shift in laboratory outlook or capabilities and provide a high degree of reversibility.

Diversification is already more difficult, in part because of incompatibility in the styles best suited to civilian as opposed to military R&D. It is also more useful insofar as it provides a foundation, as noted above, for subsequent conversion. But it does not in itself reduce the resources going to military R&D, and one may wonder whether the political attractions of diversification are enough to cause an increase in funding for civilian R&D overall, or whether the diversification of military R&D enterprises into dual-purpose enterprises simply replaces R&D that otherwise would be done elsewhere in the civilian sector.

Reprioritization, which can entail diversification but need not do so (in that the expanded civilian R&D involved might all take place in organizations other than those that had been engaged in military R&D), can, in principle, provide many of the arms-control and confidence-building benefits generally associated with reductions in military R&D, as well as providing many of the economic and public-goods benefits associated with increases in civilian R&D; but it does not do anything for the problems of unemployment, wastage of assets, social unrest, and even proliferation potential associated with putting military R&D personnel out of work without giving them something else to do. The arms-control and confidence-building benefits may be (or seem) less durable, moreover, than those associated with real conversion, in that the military R&D assets rendered unemployed by reprioritization constitute a source of economic and political pressure to reverse this change as soon as a political opportunity presents itself.

Downsizing of military R&D taken by itself has all of the liabilities of reprioritization without the benefits of increased civilian R&D. That is, although it certainly may have arms-control and confidence-building benefits and - if it is not offset by an increase in non-R&D military spending - there may be direct benefits to the civilian economy from the freeing up of the associated economic resources, it is not desirable from the standpoint of those interested in the health and productivity of the R&D enterprise. After all, from the standpoint of the health of R&D overall, military R&D is better than none, and, as noted above, the (modest) part of military R&D that is basic research and applied research not focused on particular weapon systems may be no less valuable than the same work would be if conducted under civilian auspices. It is all too possible, moreover, that downward pressure on military R&D overall will produce reductions in basic and relatively nondirected applied military research that are, proportionately, as large as or larger than the reductions it produces in military development.

Within the category of actual conversion of military R&D – as opposed to transformation, diversification, reprioritization, and downsizing of such R&D, or conversion of military production – it is useful to make three kinds of distinctions:

(1) between (1a) cases where the customer for the R&D after conversion is the same as before and (1b) cases where the customer changes (e.g., from government to industry);

(2) between (2a) cases where the sector conducting the R&D after conversion is the same as before and (2b) cases where the sector changes (e.g., from national laboratory to industry);[9] and

(3) between (3a) cases where people after conversion do tasks very similar to what they did before, but with different ends (such as developing improved gas turbines for civilian as opposed to military aircraft), hence requiring little or no retraining, and (3b) cases where the tasks or the applications are so different as to require substantial retraining (e.g., from modelling energy flows in nuclear explosions to modelling energy flows in the atmosphere).

In all three kinds of distinctions as presented here, the 'a' variants will tend to be easier to implement than the 'b' variants.

Under the heading of 'definitions' it is necessary to mention, finally, the definitional and conceptual complications introduced by the existence of 'dual-use' technologies and corresponding forms of R&D. The term 'dual use,' of course, refers to R&D activities (and products of these) that are about as likely to find application in the military sector as in the civilian sector. There is a general consensus that the possibility of dual use is greatest for 'generic' technologies – for example, new materials or information technologies – that have not yet been specialized to particular applications. There is also scope for dual use of components, even where the final product is highly specialized – for example in aircraft or satellites – assuming that the institutional barriers (discussed below) can be surmounted. Perhaps the most promising area for dual use lies with process technologies that can be adapted for either military or civilian products. Conversion issues raised by dual use are considered further below.

ASSESSING CONVERSION OPPORTUNITIES

Systematic assessment of the opportunities for conversion of military R&D requires detailed examination of the capabilities that currently exist in the military R&D sector, the civilian R&D needs that conceivably could benefit from the application of these military R&D capabilities, and the 'fit' between the military capabilities and the civilian needs. A number of assessments of this kind - for particular sectors in particular countries - are provided in this book.

Military R&D capabilities can be partially described by reference to strengths in particular disciplines and in particular military applications, wherein illustrative (certainly not comprehensive) lists would include:

Disciplines: mathematics; computing and information processing; physics; chemistry and chemical engineering; materials science; aerodynamics/fluid mechanics; electrical engineering and electronics; geology; oceanography; meteorology; biology; physiology; industrial engineering and management.

Applications: nuclear weapons design, production, and fabrication; chemical and biological weaponry and countermeasures; chemical explosives; rocket propulsion; aircraft propulsion; marine propulsion; land-vehicle propulsion; design and manufacture of satellites, missiles, aircraft, land vehicles, and ships and submarines; portable power sources; telecommunications; tracking and positioning systems; remote sensing, measurement, and imaging systems.

The civilian needs to which such military R&D capabilities might be applied can be indicated by another illustrative (rather than comprehensive) list, as follows:

Resources and Environment: energy supply, storage, and transmission; efficient energy use; water supply and quality; nonfuel materials supply; agriculture; forestry; fisheries; monitoring and understanding global, regional, and local environmental change; environmental remediation; weather prediction.

Other Industrial Technologies: manufacturing technologies; 'green' chemical synthesis and processing; recycling and waste reduction; project management.

Infrastructure: transportation systems; air-traffic control; communication systems; data-processing systems.

Human Services: medical science and technology; health-care delivery; crime prevention, detection, and response; education; recreation.

Natural-Disaster Prediction, Prevention, and Response: earthquakes; tsunamis; typhoons and hurricanes; floods; volcanic eruptions; asteroid impacts.

The key questions to be addressed about the 'fit' between military R&D capabilities and civilian needs include the following:

- How great is the relevance of the military R&D capabilities to the problematic or progress-limiting aspects of the civilian need?
- Is the scientific/technical level of the relevant military R&D capabilities greater than, equal to, or less than that of the civilian R&D capabilities already being applied to the need?
- Does the magnitude of the relevant and available military R&D resources represent a significant addition to the civilian R&D resources already being applied to the need? Can the civilian need effectively absorb the additional relevant R&D resources available to be converted from military applications?
- Who would be the client for the converted R&D activities and results? National governments? Local governments? International agencies? Corporations? What adjustments from traditional conditions and practices in military R&D would be required to match the characteristics of the new client?

Such considerations of matching capabilities and needs lead naturally to the question of barriers to conversion of military R&D, to which we now turn.

BARRIERS AND POLICIES FOR OVERCOMING THEM

Barriers to capitalizing on the opportunities for conversion of military R&D abound, and often appear to be more than a match for the incentives for conversion enumerated in the first section of this Introduction. As described in virtually every chapter in this book, these barriers include macroeconomic, political, technical/microeconomic, and institutional factors, as well as issues of monetary and psychological rewards and simple inertia.

- On the macroeconomic side, wherever and whenever high interest rates (meaning scarce capital), or high unemployment, or recession, or any combination of these prevails, conversion of R&D is impeded by the associated low incentives to invest and difficulties of job creation. This particular problematique is of course especially acute in Russia, as noted in the chapters by Maslennikov and Nikitin.
- Politically, R&D conversion is hindered by opposition of the 'iron triangle' of defense contractors, ministry of defense officials, and defense-oriented legislators, who benefit financially from military R&D and/or are strongly influenced by the defense rationales for military R&D outlined above. In addition, the high concentration of defense R&D in particular companies and regions makes for effective lobbies against conversion, as noted in the chapter by Marlin. Lack of political will by other government officials and publics to overcome this opposition arises in part from lack of knowledge about military R&D (aggravated by the secrecy and ambiguity mentioned above) or lack of conviction that high levels of military R&D spending are a problem.
- The technical/microeconomic barriers include the ingrained pattern of high-cost, high-performance design that has characterized military developments, but is not competitive in civilian markets; military specifications that are costly to maintain and irrelevant to commercial applications; and the general nonfungibility of military R&D funds into nonmilitary R&D sectors (as opposed to non-R&D military sectors), as discussed here in the chapter by Gummett and Stein.
- On the institutional side, enterprises wishing to move into new civilian markets face the need to develop new technological expertise as well as adding competence in marketing and service. This requires a degree of flexibility that does not come easily to researchers accustomed to dealing with the government customer, and in many cases it will probably require painful adjustments if not a very long process of developing a new organizational 'culture.' In addition, as noted in the chapter by Ruina, research institutions that are changing their focus – or civilian R&D institutions that are increasing their activity – may prefer to hire younger (hence lower-salaried and more recently trained) scientists and engineers rather than defense-R&D 'retreads.'
- In terms of rewards, it is frequently the case (or thought to be the case) that the professional challenges in the civilian sector are 'lower tech' and that working with them produces less prestige. Civilian R&D may also entail working in inferior facilities, for lower salaries.

- As for inertia, as long as the option of continuing in defense work remains at all feasible, many companies – and the scientists and engineers who work for them – will prefer to concentrate on the work they know best. Certainly, companies will tend to be reluctant to risk investment in a new area when resources for existing defense projects are not constrained. Thus levels of military spending that are high by historical standards, if not compared to the peaks of the 1980s, constitute in themselves an obstacle to conversion.

The barriers to conversion of military R&D are real and cannot be dismissed lightly. Nevertheless, there is reason for at least guarded optimism. The profound changes in the international situation since 1989 have persuaded decision makers both in government and in industry that military markets will be smaller for the foreseeable future. The reduction in the number of new weapons developments has necessarily put pressure on the military R&D programs that feed into them. Although initial responses were protective of established ways of doing things, with time more imaginative solutions to the cuts in military spending are appearing.[10]

Scientists in government laboratories have actively sought out new areas and have made connections to potential customers, assisted in some cases by new mechanisms created by legislation, e.g., the Cooperative Research and Development Agreements (CRADAS) used to promote lab-company cooperation in the United States. Some defense contractors have learned that a promising approach for accommodating new initiatives is to spin off the new activities as a separate division or subsidiary. Towns and regions have developed marketing fairs, information clearing-houses, and training programs.

A central question remains with respect to the appropriate and useful role for government policy. Many conversion strategies require fundamental change with regard to overcoming the 'culture' of military development and production in order to compete successfully in commercial markets. Appropriate government policy will necessarily vary with the political structures and traditions of different countries, but it is clear that governments can help. They can provide financial support for R&D for civilian-related public goods; they can act to eliminate regulatory barriers to conversion of government-operated facilities; and they can encourage partnerships between defense laboratories and civilian companies.

More controversially, governments can introduce R&D subsidies – for example through tax credits – to create incentives for companies to increase their demand for scientists and engineers and thus provide jobs for those moving out of military-related work, and they can provide investment assistance to companies wishing to enter into new markets. Certainly, Western governments could (and should) do much more than they have done so far to assist Russia in its project of converting its military industrial base to civilian production, either through bilateral programs or through support for an expanded role for the United Nations.

Finally, as noted in the chapters in this book by Marlin and Altmann et al., policies, mechanisms, and actions to help implement conversion of military R&D could and should be augmented by policies, mechanisms, and actions designed to influence society's incentives for such conversion. These might include local, regional, national, and global assessments of the civilian needs to which military R&D capabilities could usefully be applied; greatly increased transparency in military budgets (including their R&D components) to help offset the 'hedging'/worst-case assessment phenomenon that so often drives increases in this spending; and international agreements limiting military R&D spending.

THE QUESTION OF DUAL USE

Dual-use technology appears to many observers to offer a way out of the difficult problems posed by falling military budgets, rising development costs, and the challenge of international competition in civilian markets. The military technology base can benefit from the advances made in civilian technology - particularly in information technologies, where the civilian technology is in some cases more advanced than the military - and the civilian side of the economy can be boosted by improved access to the military technology base. Moreover, to the extent that technologies developed in the military sector can be identified as dual use, they form a basis for diversification and conversion of military R&D from strictly military ends; in effect, we are in categories (2a) and (3a) as described above, and we may also be in (1a), if the new customer is another branch of government. It is not surprising that policy makers have seen in dual use technology a solution to some of their most pressing problems.

Most analysts concur that there are opportunities to exploit the dual-use principle in both military and civilian applications.[11] The difficulty arises with implementing a dual-use policy and with some of the implications of such a policy if it is successful. In principle, dual use should provide a favored path to conversion. Where there is a civilian application for military-related technology, defense companies and laboratories should find it *relatively* easy to supply the demand. The main barrier is likely to be the need to develop marketing and service networks for the civilian markets; although this is a problem mainly at the production stage, it has implications for R&D, where cost-control and reliability must be given new prominence. Unfortunately, it is a serious problem, since it requires the company or laboratory essentially to adapt to a new culture.

We can add to these problems the political problems of maintaining support for a dual-use policy. In the United States, for example, while the Department of Defense has given strong rhetorical support to a dual-use policy and has initiated reductions in the number of military specifications required in defense production, Congress has reduced the funds available for dual-use programs and insisted that programs funded in the DoD be justified on military grounds only. Particular dual-use projects, such as the flat-panel display technology, have been singled out for criticism. To the extent that conversion via dual-use technology depends on government support, continued budgetary pressures in all the major R&D-performing countries threaten to undermine it.

There is another troubling aspect of government policies dedicated to promoting dual-use technology. By definition dual-use implies reversibility. Viewed from the perspective of arms control, reversibility is a problem because it undercuts the confidence building that conversion of military R&D can potentially provide to the process of mutual cuts in forces. The problem is particularly acute in countries like Germany and Japan, which have not maintained high levels of military R&D in recent years, but which have a highly developed civilian technology base. A government policy promoting dual use in these countries threatens to militarize civilian technology rather than to lead to a reduction in military technology.[12] Here we see the importance of paying attention to the varying circumstances of different countries. A policy to promote dual use may be helpful in providing a path to conversion and lower levels of military R&D in countries with large military establishments, but have the opposite effect in other countries.

CONVERSION ISSUES IN SPECIFIC COUNTRIES AND SECTORS

Our discussion so far has sought to reveal the broad outlines of conversion of military R&D, the opportunities it presents to countries in the post-Cold-War period to redirect their scientific and technological resources to civilian ends, and the obstacles that it faces in the form of entrenched institutions and habits. In practice, however, and as already noted here, the prospects for conversion need finally to be seen in their national contexts: after all, the institutions and practices of military R&D vary among countries, as do the security needs that military R&D is supposed to address. There is further variation among industrial sectors, especially with respect to the potential for following the dual-use path to conversion. The chapters in Parts II and III of this book take up these issues in considerable detail for a number of countries and sectors. Here we only point to the most important findings of those chapters.

The countries with the highest levels of military R&D, both absolutely and relative to funding for other R&D, are the United States, Russia, Britain, and France. These countries share some attributes. For example, they all have well established military-industrial establishments, including nuclear laboratories, and arguably they have all suffered in commercial markets as a result of focusing government resources for R&D on military programs. In other respects, however, these countries differ in important ways.

Russia, for example, faces unique problems as it seeks not only to convert military R&D to civilian purposes, but also to move the entire economy from state direction to market mechanisms. Military R&D occupies an uneasy position in this transition because privatization is not seen as appropriate for the core institutions of national security, but there is nevertheless a need to forge links to the private sector if conversion is to occur. The ingenuity of Russia's defense scientists in identifying potential new areas of work has not been enough in the face of the obstacles to new investment and the rapid decline in funding for civilian R&D; one result has been a substantial emigration of scientists to other countries or to other kinds of jobs in the Russian economy.[13]

The conversion task is less daunting in other countries. Although they all have experienced a degree of upheaval as a result of mergers and acquisitions in their defense industries, in no case has this upheaval been comparable to the changes in Russia.

The situation in France is marked by the important role that military R&D has played as a species of *les grands projets* and the greater willingness of the French government to engage in a proactive industrial policy, compared to the United States and Britain. It would be relatively easy from a political perspective for the French government to undertake an active conversion policy, but in fact it has shown little interest in doing so. France has cut military spending and has accepted cooperation with other European partners as an inevitable necessity, but there remains a strong core of nationalist feeling that protects military R&D as a vital source of French national power.

Britain has also cut military spending, but it has relied much more on market forces to bring about an adjustment to the new situation. The number of defense companies has been dramatically reduced through a series of acquisitions and mergers, although it is not clear that capacity has been proportionately reduced. Because British industry is largely uncompetitive in global terms, there has been an incentive for companies to remain in the defense sector where they have had relative success. In common with the other countries under discussion, arms exports have been touted as the way to maintain capacity in the industry and to justify continued investments in new developments. As in France, but for rather different reasons, the British government has had little interest in promoting conversion.

The United States is the world's largest spender on military R&D, and maintaining technological superiority continues to be at the heart of U.S. military strategy. Nevertheless, government spending on military R&D has fallen, and company spending is also falling in response to the shrinking market. There are signs that some conversion is taking place in some parts of the industry: subcontractors, vendors, and smaller companies in general have had some success in expanding their sales in commercial markets. Prime contractors, by contrast, have tended to concentrate on their core defense business, while engaging in a series of mergers and acquisitions that has substantially reduced industry capacity. Employment in defense firms has fallen, evidence that even where there has been successful conversion it has not been enough to balance the reduction in defense activities.

In the United States as in Britain, the political climate is suspicious of government policy to support conversion (or indeed any industrial development), an aversion that was particularly strong in the Reagan and Bush administrations. The Clinton administration has packaged its conversion policy under the rubric of 'technology reinvestment,' rather than orienting it to a broader set of defense activities. On the one hand this strategy has put technology, especially dual-use technology, at the center of government policy, with funding for cooperative initiatives and technology transfer that should be favorable to conversion. On the other hand, the policy is vulnerable to the criticism that it focuses on the supply of federally funded technology, with insufficient attention to the conditions necessary on the demand side to bring the technology to market. It is also problematic from the point of view of arms control, because of the reversibility feature of dual-use technology noted above.

In each of these major R&D-performing countries – and also in countries, such as Germany, in which military R&D is less important – the prospects for conversion are affected by the specific histories and institutions of the military-industrial base and beliefs about the appropriate role of government. We can also see important differences in the potential for conversion among institutions within the military-industrial base and among different industrial sectors. The presence of these differences suggests that policy should be similarly nuanced, not only in different national contexts but within a single political system, in order to accommodate the different conditions.

The national nuclear complexes, for example, are a special case in each of the nuclear powers. They represent cumulative national investments on a massive scale in a specialized technological capability. Although it is generally accepted that the nuclear laboratories are too large for the changed security situation, in each country there is a consensus around the 'need' to protect a core of nuclear-weapon competence indefinitely – an idea summed up in the U.S. phrase 'stockpile stewardship.'[14] Other current activities of the nuclear-weapon laboratories include supervising the dismantlement of nuclear weapons, improving the protection and accounting for bomb-usable nuclear materials from both the military and civilian sectors, and providing technical support for the cleanup of the damage caused by years of operating the nuclear complex without adequate regard for the environment.

Because of the special standing of nuclear-weapons technology, privatization is not an option, and the physical isolation of the laboratories in the United States and Russia and their history of secrecy set them off from the rest of the economy. While the capabilities of the laboratories might well be applied to other societal problems such as energy and the environment – and, to some extent, have been – the combination of these non-weapon-related functions and the remaining weapon-related ones has not been sufficient to absorb the considerable scientific resources that are now redundant in the weapons-development sphere. In our judgment, then, the outcome in this sector is likely to continue to be a combination of partial conversion and shrinkage, the latter by attrition to retirement, transfer to jobs in other locations, and increased unemployment.

Other sectors provide a brighter picture, but again it is necessary to distinguish among industries with respect to the importance of the military market and the opportunities presented by the current boom in information technologies. The potential for exploiting dual-use technology to make a shift from military to civilian markets is perhaps greatest in the area of information technologies, where the growth of the sector provides an expanding demand to ease the way. The importance of industry investment in R&D in the electronics and computer industries provides a degree of insulation from the effects of the decline in the defense budget.[15] The aerospace sector, however, while clearly also enjoying dual-use potential in the markets for aircraft and space vehicles, has long relied more heavily on military funding (including, in the United States, the government-funded independent R&D of the defense contractors). As a result, support for R&D in the industry has dropped sharply in the United States and Britain since the mid-1980s.[16]

The large, well established aerospace contractors have perhaps the least incentive of any sector of the defense industry to undertake conversion, since they retain a virtual monopoly on the systems capabilities that are needed for modern weapons developments. As noted above, these contractors have been active in acquisition and merger activity that has reduced the number of competitors and has led to widespread layoff of workers, but they have not been particularly interested in pursuing conversion. Looking to Russia, the situation in the aircraft industry is somewhat different because of the potential for growth in the civilian airfleet. Whether this market will be dominated by Russian production or imports remains to be seen.

CONCLUSION

The complexity of military R&D does not admit of any simple path for its economic conversion, even within a single industry or country. Individual histories, economic systems, and political cultures provide additional layers of complexity, when one looks across sectors or countries. Nevertheless, we can be confident that either conversion or downsizing will take place, if only because the combination of a changed security situation and budgetary stringency in countries around the world make further reductions in spending for military R&D inevitable. The opportunity costs of continuing to devote such large resources to military technology are not supportable in a world in which economic competition in civilian markets is the most important test of national strength for most countries.

The challenge is to find ways to make a transition to lower levels of military R&D without wasting valuable scientific and technological resources in unemployment or forced transfers to other kinds of work, and without creating new problems with investments in R&D that cloak an easy reversibility back to military ends. It is trite to remark that such policies require political will and public support, but these political assets are indeed essential if policy is to be maintained over the period needed to effect a genuine change. The institutions and culture of military R&D

were built over many years of government support, and to dismantle them will require a sustained effort.

The chapters in this book provide a thoughtful consideration of these issues and others from a variety of perspectives. They give a strong sense of what is at stake - particularly in Russia and the other nuclear successor states - in the struggle to find new uses for the military-scientific establishment and to finance the transition to new markets. The reader will see that, while there is no shortage of obstacles to wide-ranging and irreversible conversion, neither is there any shortage of ideas for new approaches and for new cooperative arrangements between countries.

NOTES

1. For world military spending at the end of the Cold War, see, e.g., U.S. Arms Control and Disarmament Agency, *World Military Expenditures and Arms Transfers 1990* (Washington, DC: U.S. Government Printing Office, November 1991). U.S. military R&D spending in 1990 was about $40 billion, and a variety of approaches to estimating the corresponding figure for the Soviet Union lead to numbers in this general range (say, US$30–60 billion). From data provided by Reppy (chapter 1, this volume), the Stockholm International Peace Research Institute, *SIPRI Yearbook 1990* (Oxford: Oxford University Press, 1990), and the International Institute for Strategic Studies, *The Military Balance 1989-90* (London: Brassey's Publishers, 1989), one may estimate that annual military R&D spending in 1989–1990 totalled about US$10 billion in the Western European countries, perhaps US$2–3 billion in the Eastern European countries, something like US$4–8 billion in China, and perhaps US$0.5–1 billion in the Middle East. The rest of the world altogether probably does not account for more than US$2–3 billion in annual military R&D (assuming, based on Reppy's figures, that in countries outside the blocs mentioned an average of 2–3 per cent of total military spending goes to R&D).
2. See, e.g., National Science Foundation, *National Patterns of R&D Resources* (Washington, DC: U.S. Government Printing Office, 1992).
3. The estimates of military R&D provided here are for spending on military R&D by governments; but some fraction of the R&D conducted by the private sector with its own funds is devoted to weapons systems, with the aim of securing government contracts to further develop and produce these, and if these private military R&D expenditures were accounted for then the military's share of world R&D spending would be somewhat higher, perhaps over 30 per cent.
4. Of course, a sensible comparison among categories of R&D spending cannot be based only the amounts spent, but must take into account the relative opportunities for productive R&D investments and the relative costs of

pursuing these. That having been said, it would be difficult for anyone familiar with the U.S. R&D scene to dispute that major R&D needs and opportunities in health, energy, environment, transportation, and agriculture, among other fields, are languishing unaddressed for lack of funds.

5. In the large literature of this subject, see, e.g., U.S. Congressional Budget Office, *Federal Support for High-Technology Industries* (Washington, DC: U.S. Government Printing Office, 1985); Kenneth Flamm, *Targeting the Computer: Government Support and International Competition* (Washington, DC: Brookings Institution, 1987); Michael E. Porter, *The Competitive Advantage of Nations* (New York: Free Press, 1990); U.S. National Academy of Engineering, *Technology and Economics* (Washington, DC: National Academy Press, 1991); John A. Alic, Lewis M. Branscomb, Harvey Brooks, Ashton B. Carter, and Gerald L. Epstein, *Beyond Spinoff: Military and Commercial Technologies in a Changing World* (Cambridge: Harvard Business School Press, 1992); Lewis M. Branscomb and Fumio Kodama, *Japanese Innovation Strategy: Technical Support for Business Visions* (Cambridge: Center for Science and International Affairs, Harvard Uniersity, 1993); Office of Science and Technology Policy, Executive Office of the President of the United States, *Science in the National Interest* (Washington, DC: U.S. Government Printing Office, August 1994); Edwin Mansfield, 'Academic Research Underlying Industrial Innovations: Sources, Characteristics, and Financing,' *The Review of Economics and Statistics* (February 1995), pp 55–65; U.S. National Research Council, *Allocating Federal Funds for Science and Technology* (Washington, DC: National Academy Press, 1995); U.S. Council on Competitiveness, *Endless Frontier, Limited Resources: U.S. R&D Policy for Competitiveness* (Washington, DC, April 1996).
6. See also the book *Beyond Spinoff* by Alic et al. cited in note 5; and Robin Cowan and Dominique Foray, 'Quandaries in the Economics of Dual Technologies and Spillovers from Military to Civilian Research and Development,' *Research Policy* 24 (1995): 851–68.
7. For example, while Japan, West Germany, and the United States spent similar fractions of their 1990 GDPs on R&D – 3.1 per cent, 2.8 per cent, and 2.7 per cent, respectively – the United States was at a substantial disadvantage in the percentage of GDP spent on nondefense R&D: 1.9 per cent, versus 2.7 per cent for Germany and 3.0 per cent for Japan. See National Science Foundation, *National Patterns of R&D Resources* (Washington, DC: U.S. Government Printing Office, 1992).
8. In the 1990 U.S. military budget, for example, the procurement account was some $80 billion and the operation and maintenance account nearly $90 billion, compared to $40 billion in the R&D account. See, e.g., Department of Commerce, *Statistical Abstract of the United States 1994* (Washington, DC: U.S. Government Printing Office, 1994). Note that these proportions vary drastically among countries owing to differences in policies, economic circumstances, and accounting conventions.

9. A case in which the sector changes can meet our strict definition of conversion – i.e., that personnel, equipment, or facilities previously used for military R&D are shifted to R&D directed at civilian aims – only if the personnel move or equipment & facilities are taken over by new management.
10. See, for example, Michael Oden, 'Cashing-in, Cashing-out and Converting: Restructuring of the Defense Industrial Base in the 1990s,' paper presented to the Council on Foreign Relations Study Group on 'Consolidation, Downsizing and Conversion in the U.S. Military Industrial Base,' New York, 9 February 1996 (mimeo); and Ann Markusen, 'The Post Cold War American Defense Industry: Options, Policies, Probable Outcomes,' in Efraim Inbar and Ben-Zion Zilberfarb, eds., *Politics and Economics of Defense Industries in a Changing World* (forthcoming).
11. For a countervailing view in relation to the electronics sector, see chapter 13.
12. See, for example, chapter 10.
13. See, e.g., Bonn International Center for Conversion, *Conversion Survey 1996: Global Disarmament, Demilitarization, and Demobilization* (Oxford University Press, New York, 1996), p. 81.
14. Of course, the admission of the nuclear-weapon states that they do not foresee any end to their asserted need to maintain nuclear-weapon stockpiles and nuclear-weapon-design capabilities is highly problematic in the context of nonproliferation policy. For an extensive discussion of the difficulties created by this contradiction, see, e.g., Joseph Rotblat, Jack Steinberger, and Bhalchandra Udgaonkar, eds., *A Nuclear-Weapon-Free World: Desirable? Feasible?* (Boulder, CO: Westview Press, 1993).
15. There are, however, some constraints on the ability of military electronics firms to take advantage of these civilian opportunities, see chapter 13.
16. National Science Foundation, *National Patterns of R&D Resources 1994* (NSF 95–304, Arlington, VA, 1995), p. 87; Directorate-General III (Industry), *The European Aerospace Industry Trading Position and Figures* (III/5021/96-EN, European Commission, Brussels, May 1996).

PART I

Setting the Context

1

Levels and Trends in International Spending for Military R&D

Judith Reppy

INTRODUCTION

How much is spent on military research and development (R&D) worldwide? In the past, much of the interest in this question focused on the role that military R&D has played in fuelling arms races, particularly the arms race between the United States and the Soviet Union. Military R&D spending in one country translates into new weapons that can feed insecurity and military spending in rival states. Moreover, the R&D-performing institutions in the major powers have been a powerful political force in favor of continued high levels of spending for new weapons development and testing, even after international tensions have eased.

The question of how much is spent for military R&D is also central to the issue of economic conversion. The amount is one measure of the opportunity cost of military R&D to society; equally, it is a measure of the size of the conversion task. Behind the budgetary figures lie the institutions and the human beings - scientists, engineers, and other technical staff - whose life histories have been devoted to designing and building weapons. When we observe that in some countries the military programs employ between a third and a half of the total scientific and technical workforce, we know that conversion to civilian purposes is both urgent and likely to be difficult.

Given the importance of the issue, it is unfortunate that there is a paucity of reliable data on military R&D spending worldwide. Some of the problems arise from government secrecy, while others are the result of inconsistency in definitions and categories between countries and of fundamental conceptual issues surrounding the distinction between military and civil R&D, problems that are discussed below. As a result of these problems we can indicate in only rough terms the magnitude of military R&D spending, although fortunately there are reasonably good data for the United States and other OECD countries.

Of equal interest is the *trend* in military R&D spending. Since the end of the Cold War military budgets have been falling in most countries, putting pressure on military R&D budgets as well. While there are incentives for governments to protect their military R&D budgets in order to keep up with the global trend to high technology weapons and to maintain their technology base, there are also demands to sacrifice R&D spending in order to maintain funding levels in categories such as training and maintenance that have a more immediate impact on military readiness. The net effect of these conflicting pressures is by no means obvious and varies among countries. The rate of decline in R&D spending has implications for future military stability as well as for economic conversion. For example, in Russia, where military spending has dropped precipitously since the breakup of the Soviet Union, there has been considerable dislocation in the research community, raising fears of proliferation through emigration of scientists and engineers. One might argue that for many policy issues, trends in military R&D spending are even more important than absolute levels, but here too we suffer from a lack of good data, especially for recent years when the cuts in military spending have been most widespread.

LEVELS AND TRENDS IN R&D SPENDING

This section presents information on levels in military R&D spending for 20 countries and trends for the six largest spenders (reliable data on China are not available). It relies heavily on recent work by the Stockholm International Peace Research Institute (SIPRI), as well as the OECD's data series, which give information on defense budget R&D as a per cent of total government R&D expenditures for most industrialized countries.[1] OECD data do not include Russia or other members of the former Warsaw Treaty Organization or important military powers in the third world such as China and India; other sources have been used to fill some of these gaps. Given the disparity of sources and definitional problems discussed in the next section, the data presented here give only a rough picture of world spending on military R&D.

The bulk of military R&D spending is concentrated in only a few countries; in most countries, particularly the less-industrialized countries, new military technology is acquired through imports, not indigenous development. The principal missions of the small programs in these countries are to keep abreast of developments elsewhere and adapt them to local conditions and to contribute to the national technology base. Pakistan, for example, has a very small military R&D budget in absolute terms - although it is a significant fraction of total government spending for R&D - but it has active programs for collaboration with China and for weapons upgrades that compensate to some degree for the weakness of the national technology base.[2] Thus, in developing countries the level of national spending for military R&D is more relevant for discussion of opportunity costs and development possibilities than it is for analyses of military capability or arms races, which are fed by arms transfers.

Table 1.1 presents data for 20 countries on the level of governmental spending for military R&D and its relationship to total government spending on R&D and total military expenditures. Spending figures are given in national currencies to avoid the problems associated with foreign exchange conversion rates. Table 1.2 presents similar data for the period 1983–94 for the six largest spenders on military R&D as a per cent of military spending. Unfortunately, no reliable information is available for China.[3] Two conclusions stand out from these tables: there is a wide disparity in the emphasis placed on military R&D in various countries, and the trend in real terms has been downward since 1990. Falling defense budgets translate into reductions in military R&D, although, with the exception of Russia and - less dramatically - Germany and Sweden, these countries have protected or increased the share of military R&D in total military spending.

The figures for Russia must be understood in the context of the chaotic nature of defense budgeting since the breakup of the Soviet Union. Inflation has been rampant, so that large increases in nominal spending translate into reductions in real terms. Projected military spending has been repeatedly adjusted downward during the fiscal year as expected revenues failed to materialize. Defense industries have continued production without orders, and payment even for firm orders has been held up by the banking system so that budget figures do not correspond to levels of activity.[4] As a result there is great confusion in the estimates of total military spending. The data for Russia in Table 1.2 are based on official defense budget figures, and indicate a sharp drop in the share devoted to military R&D within a declining total budget in real terms.

Table 1.1 **Governmental Expenditure on Military R&D**

Country	currency	year	Gov't Expend. Military R&D	as % of Goverment R&D	as % of Military Spending
Belgium	m. francs	1993	91.5	0.2	0.07
Canada	m. dollars	1991	281.3	7.0	2.19
Czechoslovakia	m. korunas	1992	657.0	–	2.40
Denmark	m. kroner	1992	35.1	0.6	0.21
France	m. francs	1991	36,002.7	37.4	14.94
Germany	m. D. Marks	1992	3,269.7	10.5	4.99
Greece	m. drachmas	1992	578.9	1.5	0.07
India	m. rupees	1993	95.9	–	4.80
Italy	b. lire	1992	856.1	7.1	2.78
Japan	b. yen	1992	125.9	5.9	2.50
Netherlands	m. guilders	1993	167.7	3.5	1.28
Norway	m. kroner	1993	428.6	5.0	1.86
Poland	b. złotys	1993	798.0	–	2.10
Portugal	m. escudos	1993	233.7	0.4	0.06
Russia	tr. roubles	1994	2.4	–	5.99
Spain	m. pesetas	1992	45,148.0	14.6	4.87
Sweden	m. kronor	1992	4,487.0	24.3	12.38
Switzerland	m. francs	1991	199.6	18.5	3.78
UK	m. pounds	1993	2,454.3	45.1	10.51
US	m. dollars	1993	41,543.6	59.2	13.97

*Not including nuclear industry.

Sources: OECD countries: OECD, *Main Science and Technology Indicators, 1993/4* (Paris: OECD, 1993); Czechoslovakia, Poland: I. Anthony, ed., 'The Future of the Defence Industries in Central and Eastern Europe,' SIPRI Research Report. No. 7; India: P.R. Chari, 'Budget 93. Smaller Bang per Buck,' *Economic Times*, 3 March 1993; Russia: SIPRI, *SIPRI Yearbook 1995* (New York: Oxford University Press, 1995), p. 406.

Table 1.2 **Trends in Military R&D as % Military Expenditures,**

Country	1983	1988	1990	1992	1994
France	11.1	11.0	13.0	11.0	10.0[b]
Germany[a]	3.2	3.5	3.8	3.7	3.5
Soviet Union/Russia	–	19.8	18.6	7.2[b]	5.6
Sweden	–	7.5	6.9	8.0	6.1
United Kingdom	–	8.8	9.0	8.6	9.7
United States	–	14.0	13.0	13.0	13.0

[a] FRG up to and including 1990 [b] 1993

Sources, OECD Countries: 1983: SIPRI, *SIPRI Yearbook 1984* (New York: Oxford University Press, 1995), Table 6.5. 1988–1994: SIPRI, *SIPRI Yearbook 1996* (New York: Oxford University Press, 1995), Table 9.3; Soviet Union/Russia: SIPRI, *SIPRI Yearbook 1994* (New York: Oxford University Press, 1995), p. 426; *SIPRI Yearbook 1995*, p. 400.

DATA PROBLEMS

Russia presents a special problem, but the data for other countries are also questionable. They suffer to a greater or lesser degree from the problems discussed below.

Secrecy

Governments often wish to conceal the size of their programs of military R&D, both for reasons of national security and to avoid domestic political criticism. Some countries go so far as to report no figure at all; China's official military budget includes neither weapons procurement nor all of R&D, and estimates of the total spent must take into account the fact that the Chinese military have access to enterprise funds in addition to the amounts budgeted by the national government.[5] Even countries that do report detailed data on military R&D routinely conceal some programs. During the Reagan administration over 20 per cent of the total U.S. military R&D budget was 'black,' and the figure for the Air Force R&D budget, which contains most of the classified space surveillance and intelligence programs, went as high as 40 per cent.[6] This lack of transparency serves not only to amplify other countries' concerns by encouraging 'worst case' planning scenarios, but also to distort the information available to those responsible for planning conversion. Where there is little or no information about the level of spending or how it is allocated among the institutions that perform military R&D, the scale of conversion needs can only be a guess. Even where the total size of secret programs, if not their details, is known, there is an impediment to conversion because the barriers erected in the name of secrecy inhibit the transfer of resources to civilian uses. Scientists and engineers may have no unclassified record of their accomplishments, and technologies that may have civilian applications are unknown to the people making investment decisions.

Definitions and Coverage

Secrecy is not the only problem in arriving at a reasonable set of estimates of world spending on military R&D. As with any categorization, there are problems of definition, problems that affect national estimates and complicate international comparisons. 'Military,' 'research' and 'development' are all contested terms in some contexts. The activities included under the rubric 'military research and development' range from basic scientific research funded in university and government laboratories, through engineering development to production and testing of prototypes. In theory, all governments follow the definitions of research and development set out in the Frascati manual;[7] in practice, ambiguities abound. Definitional issues arise, for example, where research is both fundamental and directed towards specific ends; where engineering and testing for development shade over into early production; and where technological fixes to operating systems require further development work. Government offices responsible for collecting and reporting data on R&D spending may not be close enough to the research units to be aware of the issues involved. A recent report from the UK Ministry of Defence, for example, revealed that a review of intramural military R&D had identified roughly 15 per cent of the in-house laboratories' activities that were misclassified as R&D. After deleting these budget items, the report found that the fraction of the remaining amount classified as 'research' should be increased from 30 to 60 per cent.[8]

There are also problems with coverage. The level of military R&D is not necessarily equal to the portion of the defense budget allocated to R&D activities. Other agencies may be responsible for significant military R&D projects, e.g., nuclear and space projects. Moreover, where defense procurement is based in the private sector, company spending on research and development for military projects may not be reported as defense-related. In the United States, for example, the Independent Research and Development (IR&D) Program of the Department of Defense reimburses a prorated portion of defense contractors' costs for their in-house R&D programs, but companies may not separate out their IR&D programs when reporting company-financed R&D. Unfortunately, the Department of Defense is no longer required by Congress to report on the size of the IR&D program, although it has maintained the data series so far.[9]

Table 1.3, which gives a breakdown of U.S. spending for military R&D for FY 1992, illustrates the problem. The Department of Defense's budget for Research, Development, Testing and Evaluation (RDT&E) is only 84 per cent of national defense-related R&D, which also includes the Department of Energy's spending on atomic energy defense R&D and IR&D. Even this more detailed analysis has its problems: the amount reported for IR&D covers only major contractors, suggesting a downward bias, but the portion of IR&D that is recovered as overhead on RDT&E contracts, as opposed to procurement contracts, is double-counted in the national total. The situation is similar in other countries. Detailed estimates for German military R&D, for example, suggest that true levels of spending may be as much as 50 per cent higher than the official numbers, and French official data under-report spending by other government agencies.[10]

Table 1.3 **U.S. National Defense-Related R&D Spending: 1992 (billions of current dollars)**

DOD RDT&E		
Research	1.1	
Exploratory Development	3.0	
Advanced Technology Development	6.2	
Strategic Programs	4.5	
Tactical Programs	13.5	
Intelligence & Communication	4.6	
Defense-side Mission Support	4.5	
Total		37.4
DOE Defense R&D		2.7
IR&D with Military Relevance		
Government Reimbursed	2.3	
Unreimbursed	1.9	
Total		4.2
TOTAL		44.3

Source: National Science Board, *Science & Engineering Indicators-1993* (NSB 93-1), Washington, D.C.: US Government Printing Office, 1993, p. 115.

Differences in institutional arrangements between countries render many standard comparative measures more or less useless. For example, we might be interested in the proportion of governmental spending for R&D that is allocated to military R&D, both as an indicator of national priorities and as a measure of relative technological strength. Thanks to the OECD, statistics on government spending on R&D by policy objective are available for most industrialized countries (see Table 1.1). But, while comparisons of the ratio of military R&D to all government-funded R&D over time for a single country make sense, comparisons with other countries have to take into account differing traditions of government support for R&D, which ranges from over 60 per cent of total national R&D in countries such as New Zealand, Turkey, and Iceland to 18 per cent in Japan.[11] International comparisons of military R&D as a per cent of total (public and private) R&D avoid this problem, but offer no information on governmental priorities.

Underlying Conceptual Issues

In addition to problems of definition and coverage there are underlying conceptual issues that are fundamental to the effort to get good estimates of spending on military R&D. Making the distinction between military and civilian R&D is perhaps the most serious conceptual issue from the point of view of assembling statistics that are useful for conversion policy. The interest in converting military R&D - whether driven by a concern for the arms race, dismay at the opportunity costs of military activities, or a desire to foster economic conversion - depends on the existence of a well-defined category of military, as opposed to civil, R&D. There has always been some fuzziness about these categories, but few doubted that in modern times most major military powers have had a distinct military technology base, nourished by government spending on military R&D programs and characterized by specialized institutions and products. To the extent, however, that current policies in the United States and elsewhere that are directed at increasing substantially the use of civilian technology in military programs succeed, the line between the civilian and military technology bases will disappear. 'Civilian' R&D may also be directed towards military applications while 'military' R&D may be justified on the basis of its potential and actual civilian uses.

The blurring of boundaries between military and civilian R&D has ambiguous implications for economic conversion. The increased emphasis on dual-use technologies will presumably lower the barriers to diversification into civilian markets that have effectively blocked conversion in the past. Laboratory managers will have greater incentives to seek out civilian customers and to build the kind of informal networks with civilian companies that they have so successfully constructed within the military industrial complex. Other outcomes, however, are possible. At one extreme, programs such as that in the United States to finance dual-use technology through the defense budget may end in simply producing more military technology and little conversion; the head of DOD's Advanced Project Research Agency, for example, has repeatedly asserted the primacy of military applications in the dual-use programs that he administers.[12] On the other hand, successful conversion of military R&D under the rubric of dual-use technology - that is, the replacement of military R&D programs with programs catering primarily to civilian markets - carries with it the potential for substantial, and perhaps unexamined, increases in military technological capability. Public oversight and control of military R&D will become more difficult to the extent that the center of gravity shifts from government to the private sector. These concerns must, however, be balanced against the utility of the dual-use approach in providing a path away from the dedicated military R&D programs towards increased emphasis on civilian technologies.

CONCLUSION

Data on military R&D are indispensable to policymakers and analysts interested in security, in technology policy, or in the possibilities for economic conversion. The problems of secrecy, definitions, and coverage described above require that we approach the available data with caution. Nevertheless, the data presented in this chapter allow us to draw some conclusions. They demonstrate that among the group of advanced industrialized countries, only a few – Britain, France, Russia, Sweden, and the United States – have invested heavily in military R&D. These countries have set the pace for the technological arms race (although other countries have participated through the international arms trade), and they have paid the price of diverting scientific and technological resources from civilian use to military purposes.

With the end of the Cold War, these countries face the challenge of converting institutions long accustomed to serving a military customer to activities directed towards the civilian sector. The information on recent trends in military R&D is not, however, very encouraging. With the exception of Russia, which has experienced the virtual collapse of funding mechanisms for government programs, the major countries are so far protecting their military research and development activities, even as they cut military budgets overall. The challenge to conversion advocates is to devise policies that respond to the opportunities for conversion presented by the new international situation and to provide positive incentives for deeply embedded institutions to change.

NOTES

1. Stockholm International Peace Research Institute, *SIPRI Yearbook 1996* (Oxford: Oxford University Press, 1996), Ch. 9; Organization for Economic Cooperation and Development, *Main Science and Technology Indicators, 1993/2* (Paris:OECD, 1993). For earlier studies see Randall Forsberg, *Resources Devoted to Military Research and Development* (Uppsala: Almquist & Wiksell, 1972); SIPRI, *World Armaments and Disarmament – 1987* (London: Taylor and Francis, 1987); and Council for Science and Society, *UK Military R&D* (Oxford: Oxford University Press, 1986).
2. See Ron Matthews, 'Country Survey IV: Pakistan,' *Defence and Peace Economics* 5, No. 4 (1994), pp. 323–26.

3. SIPRI estimates Chinese spending for military R&D lies somewhere between $0.5 billion and $1.5 billion in 1995. *SIPRI Yearbook 1996*, p. 384, note 9.
4. Ian Anthony, ed., *The Future of the Defence Industries in Central and Eastern Europe*, SIPRI Research Report No. 7 (Oxford: Oxford University Press, 1994), pp. 40–47; see also Alexander Oshegov, 'Financing of the Defense Conversion in Russia,' paper prepared for the International Fighter Project meeting, Washington, DC, November 11–20, 1994.
5. SIPRI, *SIPRI Yearbook 1994* (New York: Oxford University Press, 1994), pp. 441–48.
6. Steven Cain, 'The FY 1990/91 Defense Budget: Preliminary Analysis,' Defense Budget Project, Washington, DC (January 1989, mimeo), Table 10; Bill Sweetman, 'Big Bucks for Black Budget,' *Jane's Defence Weekly* (2 May 1992), pp. 763–67.
7. OECD, *The Measurement of Scientific and Technical Activities: Proposed Standard Practice for Surveys of Research and Experimental Development* (Frascati Manual) (Paris: OECD, 1981).
8. UK, Ministry of Defence, *Defence Statistics Bulletin*, No. 1 (1994).
9. National Science Foundation, *Science & Engineering Indicators - 1993* (NSB 93–1) (Washington, DC: U.S. Government Printing Office, 1993), p. 114, fn. 51.
10. See Chapters 8 and 9, this volume.
11. OECD, *Main Science and Technology Indicators, 1993/2* (Paris: OECD, 1994), p. 20.
12. See, e.g., Stacy Evers, 'ARPA Pursuing Dual-Use Technology,' *Aviation Week & Space Technology* 141 (17 October 1994), p. 56 (reprinted from *Aerospace Daily*).

2

Benefits and Costs of Military R&D

Philip Gummett and Josephine Anne Stein [1]

INTRODUCTION

The debate about the costs and benefits of military R&D is part of a wider debate about the relations between science and government. It can be traced back over several centuries, to Francis Bacon, who wrote in 1627 in his *New Atlantis* of what we could regard as a prototypical research and technology transfer organization, Salomon's House. The Fathers of this house were dedicated to the 'knowledge of causes, and secret motions of things; and to the enlarging of the bounds of human empire, to the effecting of all things possible.' But they also took it upon themselves to decide which of their inventions and discoveries should be published, and which kept secret, adding, darkly, that 'some of those we do reveal sometimes to the State, and some not.'[2]

In our own century, war, or the preparation for it, has been a great stimulant to science-government relations, in pursuit of what has always been the primary 'benefit,' or at any rate, aim, of military R&D, namely, national security.[3] Since the Second World War, military R&D has consumed the greater part of government spending on science and technology globally, and especially in the United States, former Soviet Union, Britain and France, with profound but still inadequately understood consequences for technological advance.[4]

Although the First World War saw the establishment in various countries of new governmental institutions, devoted to the advance of science and technology, it was with the build-up to, and outbreak of, the Second World War that the relationship between science and government became firmly cemented. Scientists, and other specialists, were integrated into government administration and planning, especially in the United States and Britain, in ways that they had not been previously.

They contributed more than simply inventions - for example, polythene, radar, jet engines, antibiotics, computers, and atomic energy - important though these were. They also developed new analytical tools, giving rise to the field of operations research, as they sought to apply statistical and early systems analysis techniques to military operations.

The inventions just listed, and the analytical approaches that were introduced by scientists into government (and industrial) practice, would be on any list of the non-military 'benefits' (mixed benefits in some cases, perhaps), arising from military R&D. For each one listed, critics could give counter-instances of military R&D that led to no civilian application. The problems involved in trying to achieve an analysis of costs and benefits in this area are the prime subject of this chapter. We address four main subjects, namely:

1. the effects of general defense demand on technological advance;
2. studies of spin-off and technology transfer, including:
    a) firm-level studies; and
    b) macro-economic studies;
3. manpower and training implications; and
4. security threats from military R&D.

Under the last of these, we consider the costs and benefits of military R&D in terms not of an economic calculus, but of a security one, reaching the conclusion that the relations between technologies of civil and military origin need to be rethought to reflect contemporary security, budgetary and technological conditions.

A final introductory note: in discussing this subject, we refer to institutions which perform R&D (military and civil). The nature of these, their distribution between public and private sectors, the range of work undertaken in each, and the character of the problems involved in transferring people and ideas around within a national innovation system, vary considerably between countries. This point needs to be borne in mind, and allowance made for it on occasion.

THE EFFECTS OF DEFENSE DEMAND ON TECHNOLOGICAL ADVANCE

Throughout history, defense requirements have been an important stimulus (though by no means the only one) to technological advance, particularly in times of war. The cases listed above, taken from the Second World War, illustrate the point, though we should note that not all of them *originated* in military programs, even though they were accelerated by military demand.

In fact, the relationship between technologies developed for civil and for military purposes, both in their inception and their later life, is complex. It has been particularly fully analyzed for the United States, and that body of analyses provides a useful context for our consideration of the subject here. The U.S. experience has, in important respects, provided the foundation myth, or archetype, for discussions of this subject, regardless of the different conditions that may apply elsewhere.

During and after the Second World War, the U.S. military became one of the most significant drivers of technology development in the world, both as a sponsor of R&D and as a customer for high-technology products. One reason why U.S. industry flourished immediately after the war was that military-supported technologies such as radar, electronics, computers and aircraft were applicable to the civil sector and were readily available for exploitation. In addition, many of the scientists and engineers who had worked on the wartime programs returned to civilian work, bearing their expertise with them. U.S. government investment in military R&D came to be seen as an important contributor to economic growth through commercial spin-off.

The principal mechanism for this contribution was what Ergas, writing in general terms (i.e., not just about the United States) has called the 'mission' approach to technology policy, or what the French call 'grands projets.'[5] The characteristics of 'grands projets' that have been important in this regard are that they are large in scale, long in committed time, and both exacting and highly mission-specific in their technical requirements. Within such an environment, the normal commercial criterion of return on investment is to some extent suspended. These projects require resources of a magnitude that only a government, or an alliance, can provide. On the whole, they have in practice been military in nature. Thus, Louis Gallois, then chief executive officer of the French aero-engine manufacturer SNECMA has said that 'le militaire est à l'origine de notre métier . . . Il n'y a de grands constructeurs de moteurs d'avions que dans les pays qui ont de grands programs militaires . . . [La technologie militaire] c'est la Formule 1 des moteurs d'avions.'[6]

However, despite continued high levels of military R&D spending, U.S. leadership in civil high-technology industry came under unprecedented challenges from foreign competitors in the 1970s and 1980s. The conventional wisdom that military R&D spending automatically benefited commercial industry was increasingly questioned. An appreciation grew that, because of a number of distinctions between operational requirements and institutional structures in the defense and civil sectors, and because of the need for military secrecy, separate pathways for military and commercial R&D had developed in the United States (as elsewhere). Also, the emphasis of military R&D had gradually shifted from its early postwar support for basic research, to applied research and development geared to specific military ends. At the same time, the growth of commercial markets for high-technology consumer goods, especially in electronics, provided a new impetus for technological advance that began to outstrip in importance the military impetus.

The result is that the unidirectional model of spin-off (from military to civil) is inadequate to describe the reality today. There are, for example, a number of cases of parallel technology development such as in the aircraft industry and in telecommunications, involving mutual exchange of technology between civil and military sectors. In other cases, such as advanced micro-electronics and software engineering, driven primarily by consumer and commercial market demand, civil technology development has outpaced that of the military, leading to 'spin-in' from civil to military applications. In still other cases, such as nuclear weapons or stealth technology, it is hard to envisage any civil application at all.

There are cases, especially in electronics, of technologies that originated in the civil sector but were commercialized with military support. For example, transistors were invented by AT&T, a telecommunications company. The integrated circuit was invented in 1954 by Texas Instruments, which then produced equipment for monitoring oil wells.[7] Planar processing, which enabled the mass production of silicon chips, was developed by Fairchild Semiconductors. The first metal-oxide semiconductor chip was invented by Mostek in 1969, and Intel produced the first microprocessor in 1972. All these inventions emerged entirely with internal company R&D funds.[8] But in all cases the military became an important early customer, enabling engineering development to proceed and production to begin in conditions of low financial risk.[9]

When the Pentagon was the chief customer for semiconductor products, the industry was oriented towards military requirements. But by the mid-1980s, the military share of the semiconductor market had fallen to only about 7 per cent of what the industry produced. Semiconductor R&D has become increasingly a matter for corporate finance, and accordingly has adapted to consumer demand. The state of the art of commercial semiconductor technology has surpassed that of the military by an estimated five to eight years. This has had two important consequences. First, military electronic equipment has become increasingly obsolescent relative to civil. Second, military equipment increasingly contains semiconductors made outside the United States. Much the same story could be told for the U.S. software market.[10]

Thus the history of relations between civil and military technologies in the United States since 1945 suggests that the impetus that all wars give to technical advance continued for a decade or so, but began to peter out as civil demand grew, especially in electronics. At the same time, the military requirement for certain products, such as supersonic, highly agile aircraft, or precision inertial guidance,[11] diverged to the point where they had little in common with their civil counterparts, further reducing the scope for benefit to the civil economy from military investments in technology.

By the late 1980s, the military spin-off model was seen as an inadequate basis for managing the scientific and technological resources of the United States. Viewed in global perspective, whereas in 1960 the U.S. Department of Defense funded one-third of all R&D in the western world, by the 1990s it funded less than one-seventh. Although U.S. defense R&D spending had increased by about 50 per cent in real terms during this period, its significance had fallen because R&D spending by the rest of the western world had gone up by a factor greater than five.[12]

STUDIES OF SPIN-OFF AND TECHNOLOGY TRANSFER[13]

Another way of approaching the subject is to consider attempts at collecting systematic data on spin-off and transfer of technology from programs of military origin to civil application.

Much of the evidence relating to spin-off in the UK, as in the United States, is anecdotal, based on individual cases. For instance, in 1986, the Ministry of Defence supplied a parliamentary committee with a list of important spin-offs from its defense research establishments. The list included improvements in helicopter rotor systems, the development of new aluminum-lithium alloys, and research into methods of reducing aircraft engine noise, each of which had been of value to civil as well as defense users.[14] Other well-known cases include liquid crystal displays, equipment for growing semiconductor crystals, various items of computer software, pyroelectric devices (for use by fire services and in medical applications), photon correlation spectroscopy (used in applications that vary from measuring retinal blood flow to improving design of diesel and jet engines), lasers, gallium arsenide devices, and X-ray scintillators.

The drawback with this type of evidence, however, is that it can be misleading when used as the basis for generalizations. Individual cases of spin-off tell us little about the overall picture, particularly the rate of spin-off, and how it compares with the rate of successful innovation in the civil sector. Hence the interest in more systematic firm-level surveys, and larger-scale macro-economic analyses, which have been attempted in a number of UK and American studies.

Firm-level Studies

One example of such a study was the evidence, based on a survey of its members, given by the UK Defence Manufacturers' Association in 1986 to a parliamentary committee. This stated that 37 out of 142 respondents had claimed that they had achieved a spin-off from their defense work to civil interests.[15]

A similar study was performed in 1986–87 by General Technology Systems Ltd for the UK Department of Trade and Industry.[16] The study surveyed 40 defense contractors who between them accounted for £5 billion of the Ministry of Defence's 1985–86 procurement budget of about £9 billion. The companies reported about 160 product lines which originated in defense technology. These produced a civil turnover of £3.2 billion per year, which represented about 20 per cent of the civil sales of the 40 companies.

In both cases, however, evaluating these studies is difficult. Without a 'benchmark' for the rate of technology transfer between one *civil* activity and another, it is impossible to say whether the data from these surveys indicate a good level of transfer from *defense* work to civil. Moreover, the information published about the General Technology Systems survey (only a summary has been published, not the full report) leaves key questions unanswered. We do not know, for example, whether the criteria that were used to identify technologies of defense origin rigorously excluded technologies that had received some civil funding. In other words, we do not know how 'pure' are the cases of defense-originated technologies. Secondly, the study shows that, on average, each of the companies surveyed could identify four civil product lines of military origin. Whether that is a good or a poor return depends on, among other things, the time period over which these products emerged, on which we are told nothing.

Another point that emerged from the General Technology Systems report is that defense R&D varies widely in its suitability for adaptation to civilian uses. In this vein, a 1989 report by the UK Cabinet's Advisory Council for Science and Technology (ACOST) concluded that less than 20 per cent of the British government's defense R&D budget was aimed at technologies that were likely to generate civil spin-offs, while the German Ministry of Science and Technology in 1990 estimated spin-off from German firms engaged in military programs at 10–15 per cent at best.[17] Much defense R&D involves the development of existing technology through its application to very specific requirements. In fact it is far from easy to distinguish the final stages of development, such as the construction of prototypes, from mainstream product engineering. This kind of work is also usually too specialized to be of much relevance to the civil sector. In general, the main opportunities for spin-off at the development stage are likely to be in the aircraft industry, rather than in warships, tanks, ordnance or munitions, and even here there are limits to what is plausible. The requirements for warfighting aircraft of maneuverability, speed and stealth, are quite different in character from those for commercial airliners, which value stability and affordable operating costs, and must be conspicuous to radar! Similar points had been made in 1983 by the Maddock report, which emphasized also the way in which differences in company size, markets and internal structure could influence the propensity towards spin-off.[18]

Macro-economic Studies

With studies based on firm surveys proving so unhelpful, what can be gained instead from attempts to discern a relationship between national spending on defense R&D and overall national economic performance?

Such studies, it should be said at the outset, suffer from two great difficulties: first, the doubtful quality of data relating to military R&D,[19] and second, our limited understanding of the relations between R&D inputs, and economic outputs - an understanding which falls far short of what is needed to model those relations econometrically in any rigorous way. Indeed, some of these studies plot such simple correlations between R&D expenditure and some indicator of economic performance (such as productivity growth), without any attempt at specifying intermediate variables, or any attempt to allow for the effects on the analysis of the outlying cases (typically the former USSR at one extreme and Japan at the other), that their value is immediately reduced.[20]

Hartley has attempted to test the hypothesis that expenditure on defense R&D crowds out investment by looking at data for some major OECD countries between 1961 and 1987.[21] The results were far from conclusive. In Germany, defense R&D appeared to crowd out investment, but in the UK it appeared to encourage it. In several other countries (the United States, France, Japan and Holland) the direction of causation was unclear.

In the United States, statistical data allow analyses that in some respects are impossible elsewhere. First, there have been several studies into the economic effects of federal government R&D. Although this work failed to distinguish between civil and military funding, the overwhelming domination of the federal R&D budget by defense interests makes the studies relevant to our question. Griliches edited a volume of papers on this issue in which J.T. Scott found that government funded industrial R&D in the United States stimulated private R&D, and Mansfield came to a similar conclusion, but Griliches and Lichtenberg could discover no significant relationship between government funded R&D and total factor productivity.[22] A later paper by Lichtenberg found that the overall effect of federal industrial R&D was to reduce private industrial R&D.[23]

Yet another paper by Lichtenberg, however, using particularly detailed Census Bureau data that allowed a distinction to be made between types of R&D contracts at the level of individual firms, found that when federal R&D contracts were *competitively* awarded, firms' own spending on R&D went up; and vice versa.[24] This interesting finding suggests that it is the conditions under which the contract is won, therefore, rather than whether it is from a civil or a military agency, that stimulates or diminishes firms' own R&D investment, it being assumed that firms' own spending is strongly linked to their best judgment of commercial possibilities.

Another approach, developed by Reppy, attempts to test whether there has been a difference between countries in the export performance of high technology goods associated with the military technology base as compared to civilian-based goods. An initial study, comparing the United States and Japan, found that in the period 1963–79, the United States lost market share in seven of the ten product groups identified; and that Japan gained ground in most of the ten. But so many difficulties attached to the construction and interpretation of the data that Reppy argued that the easy conclusion – that high spending on military R&D leads to disadvantage in exports of civilian high technology products – should be distrusted.[25] In a later study, covering more countries, Reppy again concluded that the question remained open.[26]

These studies, then, leave open the question of whether defense R&D crowds out investment in civil R&D, or indeed leads in other ways to wasteful use of public resources. The evidence is unclear, but certainly it does not lead, *prima facie*, to the conclusion that any reduction in military R&D spending will be economically significant beyond its direct impact on the researchers involved. We should also note the frequently heard argument that such reductions would free funds for allocation to civil R&D, thus aiding the conversion process by creating new demand for displaced scientists and engineers. In practice, however, government budgetary procedures rarely work like that. Military R&D funding is usually perceived as part of the overall military budget, not as part of an overall science and technology budget. Hence reductions in it would be more likely to result in alternative use of defense funds (or even budget reductions) than to increased civil R&D spending. Any attempt to reserve those funds for conversion would have to be fought for politically.

MANPOWER AND TRAINING IMPLICATIONS

One area, however, in which it seems more plausible to argue that military demand might be constraining civil supply is in the field of manpower. Certainly, it has often been argued that defense R&D pre-empts the supply of skilled manpower. Nelson wrote of defense R&D in the United States 'taking the cream of the new science and engineering graduates.'[27] Kaldor et al. claimed, with respect to the UK: 'The absorption of a large proportion of the pool of skilled manpower by military R&D and production is arguably the most significant cost to the British industrial economy, especially since Britain is by common consent an under-skilled society.'[28]

To establish the case for crowding out, it is not enough to show that there are large numbers of qualified scientists and engineers employed on defense work, or that a high proportion of the annual output of, say, information technology graduates goes into defense work. Unless there is also evidence of unsatisfied demand for these scientists and engineers in civil industry (such as real job vacancies for which candidates cannot be found), the existence of crowding out cannot be demonstrated. With limited exceptions, such evidence is hard to find, though this could be a function of the poor quality of manpower statistics rather than necessarily an absence of the phenomenon itself.

Thus, in the case of the UK, official manpower statistics do not help with this question. The most reliable evidence concerns electronics and information technology graduates, of whom a special study was done for the Cabinet's Advisory Council on Science and Technology for its 1989 report on defense R&D.[29] The study estimated that in 1987 there were 230,000 scientists and engineers in electronics and information technology, of whom 20–25,000 (7–9 per cent of the total stock) were employed in work funded by MOD. In the United States the proportion of such scientists and engineers working in defense was much higher, between 10 and 20 per cent. The study included a survey of five major electronics firms and three software houses, finding that they recruited

1,400 information technology graduates (about one-quarter of all new electronics and computer graduates) during 1987 for defense projects. In another study, Kaldor et al. found that in GEC (the UK General Electric Company), the proportion of scientists and engineers to all employees in civil and defense work were 14 and 32 per cent respectively.[30] These figures suggest that defense R&D does have a very significant impact on the market for scientists and engineers, but the nature of that impact is unclear, given the dynamic character of manpower markets. That is, it is unclear how many students would have undertaken training in information technology subjects, and other subjects drawn upon by the defense market, had they not felt confident of employment prospects, created in part by defense demand.

In the United States there is some evidence to suggest that the employment of scientists and engineers in defense work forces up wage levels. Freeman examined the market for physicists and engineers between the 1940s and 1970s. He found that changes in overall federal R&D spending (civil as well as defense) was the primary determinant of changes in physicists' salaries.[31] In the UK, on the other hand, there are no data on this score, but also no evidence that recruitment to the civil sector has suffered from a drain of skilled staff to the defense sector. Similarly, in the United States, a study for the National Academy of Engineering concluded, first, that despite the sharp increase in defense outlays under the Reagan presidency in the 1980s, there was 'no evidence that the defense program seriously depleted the supply of Ph.D.-level engineers and scientists available for nondefense work,' and second, on the question of whether defense activity was disproportionately attracting the brightest and best, that 'data were inadequate to address the hypothesis . . . there is no evidence at present that either the defense sector or the commercial sector is lacking in appropriate representation of the 'best' of our engineers.'[32]

What can we conclude on the manpower question? On the evidence available, it is not clear that the defense sector has drained staff away from civil demand. Defense demand itself has probably increased the supply of scientists and engineers. There is no clear evidence of widespread unfulfilled demand for scientists and engineers for civil work. In respect of conversion of military R&D, therefore, the question must be put in terms of the creation of new demand for displaced military R&D staff, rather than of any easy assumption that there is a civil labor market waiting to absorb them.

SECURITY BENEFITS AND THREATS

The ostensible purpose of military research and development is to underpin national security. To the extent that it does this, then it is satisfying its primary goal, and the search for evidence of other (e.g., economic) benefits becomes a secondary matter.

On the other hand, the argument is often made that military research and development has been the motor of the arms race, with countries vying to outdo each other in advanced weaponry so as to avert the threat that other countries might steal a technological march on them at some time in the future. Just such arguments are being used at present by, for example, the nations capable of producing fighter and strike aircraft, with each claiming that aircraft being developed for export by others provides the benchmark against which their own equipment must be measured.

A variant of this argument suggests that military technology is somehow out of control, driven by its own 'mad momentum.' The argument is too well known to require footnoting. Against it, however, can be put the view that arms races occur in response to conditions of insecurity, and that it is only in such conditions (real or imagined) that the funds needed for military research and development become available. Current conditions, therefore, with the end of the bipolar Cold War, would appear to offer an opportunity for cuts in military R&D, with consequent reductions in the rate of military technological change.

Changes in the way that defense technology may be developed could transform the terms of the debate above, though not in a simple way. In brief, we may today be on the verge of a break away from the post-1945 assumptions that military technology is inevitably superior to civil, and that it must be pursued through specialist institutions. Instead, we may be returning to the more 'normal' position of the late 19th and early 20th century, in which military products developed from a strong general technology base in, in particular, Britain, Germany and the United States.

Pressures to move in this direction arise partly from the budgetary constraints that defense ministries have to face, coupled with demands on them for ever more sophisticated weapons (characterized in particular by smartness and accuracy, the former reducing the risk to one's own forces, and the latter reducing the risk of 'collateral damage'). But there is another important aspect of the defense scene that needs to be addressed. Increasingly, the prime contractor, at the top of a hierarchy of subcontractors, functions as design and management authority, systems integrator and final assembler, importing most of the sub-assemblies from its subcontractors. The complexity of the technologies that make up sub-assemblies means that the fraction of value-added by the suppliers is growing, increasing the dependence of the prime on its suppliers and squeezing its profit margins (to which, of course, the primes respond through various strategies, one of which is vertical integration). Just such a concern was voiced by the chairman of Vickers Defense Systems (the British tank manufacturer) who warned in May 1990 that companies faced a range of threats, 'not only from their direct competitors but also from suppliers developing technologies that form the core of the final product.'[33] This is particularly an issue in electronics, where economies of scale give an increasing advantage to the major civil suppliers over small defense suppliers, and where it has already been accepted for some years that, in the field of Very Large Scale Integration, military electronics must use civil components, albeit in modified form. As budgetary pressures grow, the appeal of defense equipment that is developed on the back of civil technologies will also grow. So too will the scope for civilian holders of technology to move sideways into defense markets. To a degree this is already happening, as shown recently by IBM's success in winning the contract to manage the electronics package in the European EH101 medium-lift helicopter project.

These developments put a premium on the mastery of so-called 'dual use' technologies, that is, technologies that can serve both military and civil ends.[34] They also suggest a way of satisfying the needs of defense ministries for developing advanced technologies at a time when they are decreasingly able to pay for them (though vigilance should be maintained to ensure that the label 'dual use' is not being used simply to disguise continued investment intended purely for military purposes). They could do this in particular by helping with a third, related, problem: how to keep technologies available without necessarily going into production of new defense equipment? This is a possibility to which a former British defense procurement minister, Alan Clark referred in September 1991,[35] and of which there has been much discussion in other countries, not least the United States, where there has been a big debate not only about dual-use technologies but also about identifying and supporting so-called 'critical technologies.'[36] The trouble, however, is that firms cannot live by R&D alone. They need production for survival.

What is needed is to shift defense procurement onto something more like the 'Japanese model.' That means rooting most defense production within firms whose main business is NOT defense; and who draw on a general technology base for this purpose, not on segregated civil and military ones, as has largely been the case in the UK and the United States.[37] That technology base, suitably fostered by defense ministries as well as by other sources, could provide the capability for future defense production, while the mainly civil production of the firms could provide their principal sustenance, thus reducing the importance to them of defense business, and defense cuts. An important corollary, however, is that attempts to control the international flow of militarily-relevant technology would also be greatly complicated.[38]

These developments would be part of a wider conceptual shift in our concept of 'security.' Instead of thinking of security simply in military terms, it would instead be conceived in terms of a wider range of factors that bore on the issue of national freedom of action. In this way, the debate about the costs and benefits of military R&D would be transformed, and the problem of conversion would begin to be addressed from a new, and healthier, base.

NOTES

1. Acknowledgement: This chapter draws on work done jointly with William Walker, senior fellow at the Science Policy Research Unit, University of Sussex, funded by the Economic and Social Research Council under a joint project on European Integration in Defence Markets. It also draws on work done jointly with the ESRC/Science Policy Support Group defense science and technology policy team in the preparation of their report, *Future Relations Between Defence and Civil Science and Technology* (London: Science Policy Support Group Review Paper No. 2, 1992), written for the UK Parliamentary Office of Science and Technology.
2. A. Johnston, ed., *Francis Bacon* (London: Batsford, 1965), pp. 175–80.
3. D. Edgerton and P. Gummett, 'Science, Technology and Economics in the Twentieth Century,' in *British Military History: A Supplement to Robin Higham's Guide to the Sources*, ed. G. Jordan (New York & London: Garland Publishing, 1988).
4. See P. Gummett and J. Reppy, eds., *The Relations Between Defence and Civil Technologies* (Dordrecht: Kluwer, 1988); D. MacKenzie, 'Science and Technology Studies and the Question of the Military,' *Social Studies of Science* 16 (1986), pp. 361–71; and E. Mendelsohn, et al., eds., *Science, Technology and the Military* (Sociology of the Sciences), 2 vols. (Dordrecht: Kluwer, 1988).
5. H. Ergas, 'Does Technology Policy Matter?' OECD paper WO212H/10.7.86 (Paris: OECD, 1986) (mimeographed).
6. Interview in *Info DGA*, no. 41 (Paris, January 1992), p. 37. Translation: 'the military is the basis of our field . . . Large aero-engine manufacturers are found only in countries with large military programs . . . [Military technology] is the Formula 1 of aero-engines.'
7. J.E. Schnee, 'Government Programs and the Growth of High-technology Industries,' *Research Policy* 7 (1978).

8. B. McCarten, 'Military Research and the Economy: Burden or Benefactor?' *The Defense Monitor* 14, no. 1 (1985).
9. For further details, see ESRC/SPSG Defence Science and Technology Policy Team, *Future Relations Between Defence and Civil Science and Technology* (London: Science Policy Support Group, for Parliamentary Office of Science and Technology, SPSG Review Paper No. 2, 1991); or J. Stein, *Military R&D and Commercial Technology*, Contractor Report to U.S. Congress, Office of Technology Assessment, 23 September 1988.
10. K. Julian, 'Defense Program Pushes Microchip Frontiers,' *High Technology* (May 1985); Bruce Stokes, 'Fighting Separate Wars,' *National Journal* (14 March 1987); Defense Science Board, 'Military Software,' (Washington, DC, September 1987).
11. On inertial guidance, see D. MacKenzie, *Inventing Accuracy: A Historical Sociology of Nuclear Missile Guidance* (London: MIT Press, 1990).
12. J. Alic et al., *Beyond Spin-off: Military and Commercial Technologies in a Changing World* (Boston: Harvard Business School Press, 1992), p. 5.
13. This subsection draws heavily on work done by Keith Hartley for section 4 and appendices 3 and 4 of ESRC/SPSG, *Future Relations*.
14. House of Lords, First Report of the Select Committee on Science and Technology, Session 1986–87, *Civil Research and Development* (London: HMSO, HL 20–II), pp. 463–78.
15. Select Committee on Science and Technology, *Civil Research and Development*, pp. 92–4.
16. Department of Trade and Industry/Ministry of Defence, 'The Potential for Civil Benefit From Defence Research and Development' (London: DTI Library, 1989) (mimeographed).
17. Advisory Council on Science and Technology, *Defence R&D: A National Resource* (London: HMSO, 1989), paragraphs 3.7–3.13; and Deutscher Bundestag, *Die Entwicklung der Ausgaben für militärische Forschung und Entwicklung sowie für Friedens- und Konfliktforschung* (Response of the Federal government to a parliamentarian request by the SPD), Bundestagsdrucksache 11/7373 (Bonn, 12.6.1990), p. 20, cited in J. Altmann et al., 'Dual Use and the Conversion of Military Related R&D in Germany,' ch. 9, this volume. The German source also said that the equivalent U.S. figure was believed to be 5–6 per cent.
18. Sir I. Maddock, *Civil Exploitation of Defense Technology: Report to the Electronics Economic Development Council* (London: National Economic Development Office, 1983).
19. In the UK there has been extensive debate in recent years about, among other things, major discrepancies between the Ministry of Defense's figures on R&D performed under contract in industry, and industry's figures on the same subject. Relevant sources include: ESRC/SPSG, *Future Relations* appendix 1; National Audit Office, *Classification of Defense Research and Development Expenditure* (London: HMSO, HC 105, 1991); and D. Buck, 'Problems with UK Defense R&D Statistics,' *Science and Public Policy* 20, no. 1 (February 1993), p. 48. See also *Defense Statistics Bulletin No. 1*, 'Changes to the

Statistics of MOD's Intramural Research and Development Expenditure' (London: Defense Analytical Services Agency, Ministry of Defense, 1993) for a description of changes made to another element of the UK statistics.

20. For example, R.W. DeGrasse, Jr., *Military Expansion Economic Decline* (New York: Council on Economic Priorities, 1983), chart 3.4
21. K. Hartley and J. Singleton, 'Defense R&D and Crowding Out,' *Science and Public Policy* 17, no. 3 (1990), pp. 152–56. He chose to test this particular relationship because of the words used in 1987 by the UK Ministry of Defense when it acknowledged the possibility that 'necessary investment in defense R&D may crowd out valuable investment in the civil sector,' Ministry of Defense, *Statement on the Defense Estimates 1987* (London: HMSO, Cm101-I), para. 522.
22. Z. Griliches, *R&D, Patents, and Productivity* (Chicago and London: University of Chicago Press, 1984).
23. F. Lichtenberg, 'The Private R&D Investment Response to Federal Design and Technical Competitions,' *American Economic Review* 78 (1988), pp. 550–59.
24. F.R. Lichtenberg, 'Crowding Out: The Impact of the Strategic Defense Initiative on U.S. Civilian R&D Investment and Industrial Competitiveness,' paper given at a conference of the International Council for Science Policy Studies and Swedish Council for Planning and Coordination of Research, Malmö, Sweden, 1988. Lichtenberg's results show that a $1 rise in *competitively awarded* federal R&D contracts leads to a $0.86 rise in privately funded R&D; and that a $1 rise in *non-competitively awarded* contracts leads to a reduction of $2.11 in privately funded R&D.
25. J. Reppy, 'Technology and Trade: Does Military R&D Make a Difference?' in *The Relations Between Defense and Civil Technologies*, ed. P. Gummett and J. Reppy (Dordrecht: Kluwer, 1988).
26. J. Reppy, 'Military Research and Development and International Trade Performance,' in *The Political Economy of Defense: Issues Perspectives*, ed. A. Ross (New York, Greenwood Press, 1991).
27. R.R. Nelson, 'The Impact of Arms Reduction on Research and Development,' *American Economic Review* 53 (1963), p. 445.
28. M. Kaldor, M. Sharp and W. Walker, 'Industrial competitiveness and Britain's defense,' *Lloyds Bank Review* (October 1986), pp. 31–49.
29. ACOST, *Defense R&D*, p. 56.
30. Kaldor et al., 'Industrial Competitiveness,' p. 40.
31. R.B. Freeman, 'Supply and Salary Adjustments to the Changing Science Manpower Market: Physics, 1948–73,' *American Economic Review* 65 (1975), pp. 27–39.
32. Panel on Engineering Labor Markets, Office of Scientific and Engineering Personnel, National Research Council, *The Impact of Defense Spending on Nondefense Engineering Labor Markets: A Report to the National Academy of Engineering* (Washington, DC: National Academy Press, 1986), p. 10.
33. C. Leadbeater, 'Industry Urged to Speed up R&D,' *Financial Times* (24 May 1990).

34. See Gummett and Reppy, *The Relationships between Defense and Civil Technologies*; J. Alic et al., *Beyond Spinoff: Military and Commercial Technologies in a Changing World* (Boston, MA: Harvard Business School Press, 1992); and J. Reppy, 'Steering Military R&D,' and J. Molas and W. Walker, 'Military Innovation's Growing Reliance on Civil Technology: A New Source of Dynamism and Structural Change,' both in *Military Technological Innovation and Stability in a Changing World*, ed. W. Smit, J. Grin and L. Voronkov (Amsterdam: VU University Press, 1992).
35. A. Clark, Speech on 'Defense and the High Technology Industries' (London: World Economic Forum, 4 September 1991); available from Ministry of Defense as News Release 106/91.
36. Office of Technology Assessment, *Holding the Edge: Maintaining the Defense Technology Base* (Washington, DC: U.S. Government Printing Office, 1989).
37. These ideas, though not the expression 'the Japanese model,' are discussed in R. Drifte, *Arms Production in Japan: The Military Application of Civilian Technology* (Boulder, CO: Westview, 1986); and Richard Samuels, *Rich Nation, Strong Army: National Security and the Technological Transformation of Japan* (Ithaca, NY: Cornell University Press, 1994).
38. Though after the experience of the Iraqi nuclear program, and the scandal in the UK dating from November 1992 over the transfer of dual-use technology to Iraq, it is a moot point how effective existing controls are anyway!

PART II

Conversion of Military R&D

3

Conversion of Military R&D: Concepts and Complications

Peter Southwood and Yuri Andreev

BACKGROUND

The Context for Conversion

The term 'conversion' implies a deep change. It involves a radical transformation. When applied to military industries the idea of conversion recalls the age-old vision of a society without war in which swords would be turned into plowshares.

While the formal ending of the Cold War in November 1990 created a unique opportunity to put conversion ideas into practice, little was done about them in the West at first. More recently, the Clinton administration has taken some limited initiatives to support conversion in the United States.[1] The reasons for this inertia are deeply rooted in the structures of militaristic institutions and it is important, therefore, to study conversion in this wider context. Unless this is done only a rather simplistic notion of the meaning of, and potential for, conversion can be derived. Similarly, military research and development (R&D), while being at the heart of the technological arms race, cannot be understood aside from its relation to other military activities. Consequently, the conversion of military R&D has to be undertaken as part of the overall transformation of the war economy and military institutions to peaceful purposes.

The conversion process may be undertaken as part of either partial arms reduction measures (involving a restructuring of military industries and research establishments) or general and complete disarmament. However, some writers have criticized traditional, liberal views of conversion for failing to take account of the politics of conversion and the need to develop a view of an alternative society. A radical perspective has been advocated by Oberg who, following Kaldor, argued that:

> . . . we would hold it highly unlikely that conversion will lead to real disarmament unless it is part of a *larger strategy* which encompasses (a) a converted attitude among decision-makers as to what *security* is about, (b) a fundamental democratic control of the *social driving forces behind armament* and (c) a concern for the human and social development brought about through *the structural transformation and dissolution of those conflict formations that lead to war.*[2] [Emphasis in the original.]

This radical view of conversion is important in stressing the structural obstacles to be overcome if conversion is to contribute to disarmament rather than just being incorporated within militaristic strategies of adjustment.

So the process of conversion needs to be analyzed using a contextualized or systems approach. The first question to consider is the correlation between conversion and the military doctrine, strategy and policy of the country concerned. Only after agreement about doctrine could the country put aside those capacities needed for military R&D and production; the rest could be clearly considered as objects of conversion.

In Russia the debate on military doctrine concerns reasonable or minimal sufficiency. The result of this debate will decisively influence the limits, the entire terrain for the process of conversion of military production and R&D in Russia. However, the plan for military reform already prepared, which envisages a gradual but considerable decrease in the size of the armed forces and eventual transfer to a professional army, indicates the direction the development is taking. In other countries this correlation is also extremely important, and any development of their military doctrines in the positive direction of disarmament will necessarily increase the possibilities for conversion.

The systems approach also includes the social aspects of the process. In both the United States and Russia these aspects draw special attention from the government. In Russia a considerable part of the workforce is employed in the military-industrial complex; their position and social mood are very important for the survival of any Russian government. In the United States the situation is not so acute, but in many aspects it is quite similar: the lawmakers and the executive branch are always paying attention to the social situation in this area. Thus programs of conversion in any country must incorporate in the plan of possible actions answers to social challenges: re-education of the workforce, social insurance schemes, and so on.

The Relationship Between Military R&D and Other Military Activities

The context of military R&D activities is a significant aspect of conversion and needs to be briefly summarized. Military R&D is part of the procurement process of major national defense ministries for the domestic supply and export of weapons systems. This process is an immensely complex one which involves political, military, technological, financial and industrial decisions. Figure 3.1 provides a simplified illustration of the procurement process, using the UK as an example. Similar charts could be drawn for the other major arms-producing countries in the West. It will be seen that military R&D relates to stages 1–4 and clearly lies at the heart of decision-making on military equipment purchases.

Increasingly in Europe these decisions have involved collaboration through international institutions such as the North Atlantic Treaty Organization (NATO), the Independent European Programme Group (IEPG) and the European Co-operation Long-Term Initiative for Defense (EUCLID). The limited extent of accountability at both national and inter-governmental levels over an area which by its nature tends to be very secretive has led various non-governmental organizations to express their concern.[3]

The close links between military R&D and the rest of the procurement process, which in turn are related to issues of military doctrine and security, emphasize the inter-dependency of conversion at all levels. A major program of military R&D conversion could not be carried out in isolation. The factors which might influence the success or failure of conversion initiatives, the extent of military secrecy being one example, will be examined later. The first step, though, is to consider the relationship between military and civilian R&D, which is the starting point for the conversion of military R&D.

The Relationship Between Military and Civil Science and Technology

A recent report from the UK Parliamentary Office of Science and Technology (POST) stressed that there is nothing fundamental to distinguish many areas of science and technology as being civil- or military-oriented.[4] This is especially so for 'generic' research and enabling technologies. Divergence between the two sectors is far more in evidence at the development stage of the R&D process.

Figure **3.1**
Stages in the Life Cycle of Major R&D Programs in the UK

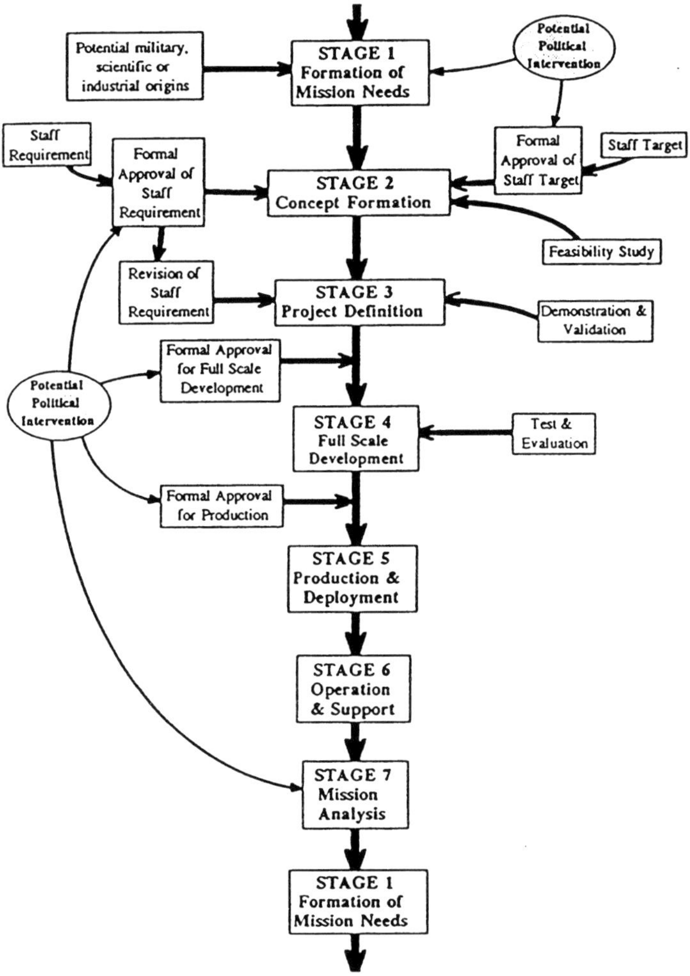

Source: Ian Davis, *Military R&D in Europe: Collaboration Without Control?*, Current Decisions Report No. 11 (Oxford: Oxford Research Group, October 1992), p. 40.

Historically, Western military R&D has led to substantial developments in the civilian sector, or 'spin-off,' as it is often called. Increasingly, though, the technology is advancing more rapidly in the civil sector, with opportunities for military 'spin-on,' especially in the fields of electronics and materials. Consequently many technological areas are potentially 'dual-use.'[5]

Nevertheless, this concept of dual-use is a problematic one. In the United States much effort has been expended in identifying 'critical technologies' seen as essential prerequisites for economic and military strength. Table 3.1 lists such technologies from the viewpoint of commercial competitiveness and military requirements. By comparing the lists it appears that many fields of technology are critical to civil and military needs. However, 'dual-use policy which focuses on core capabilities in electronics, microelectronics, computing, software and materials could easily lead to greater involvement of civil industries in defence contracting rather than vice versa.'[6] Such a policy is being used as part of a political campaign in the United States to sustain rather than convert military R&D resources.

Underpinning the growing significance of dual-use or generic technologies for developments in the civil and military arenas is the argument that economic and military strength is dependent on the national technology and industrial base. The U.S. Office of Technology Assessment has illustrated this, as shown in Figure 3.2, with the analogy of a tree, in which the 'roots' are generic technologies, the 'trunk' a country's technology and industrial base, and the 'branches' increasingly specific applications.[7] The extent of commonality between military and civil technologies should not be exaggerated, however, as differences in application arise between the two sectors due to varying market demands, and for administrative and institutional reasons. Among the latter considerations are business practices, government contracting procedures and bureaucracy.[8]

So, in general terms, the conversion of military R&D is likely to be easier in those institutions which concentrate on developing dual-use technologies in the early stages of the procurement process, but the existence of such technologies raises important questions about the irreversibility of the conversion process.

Table 3.1
Comparison of National Critical Technologies with Department of Commerce Emerging Technologies and Department of Defense Critical Technologies in the US

National Critical Technologies[a]	Commerce Emerging Technologies[b]	Defense Critical Technologies[c]
Materials		
Materials synthesis and processing	Advanced materials	Composite materials
Electronic and photonic materials	Advanced semiconductor devices	Semiconductor materials and micro-electronic circuits
	Superconductors	Superconductors
Ceramics	Advanced materials	
Composites		
High-performance metals and alloys		
Information and Communications		
Software	High-performance computing	Software producibility
Microelectronics and opto-electronics	Advanced semiconductor devices	Semiconductor materials and micro-electronic circuits
	Optoelectronics	Photonics
High-performance computing and networking	High-performance computing	Parallel computer architectures Data fusion
High-definition imaging and displays	Digital imaging	Signal processing
Sensors and signal processing	Sensor technology	Passive sensors
Data storage and peripherals		Sensitive radars
Computer simulation and modeling	High-density data storage High-performance computing	Machine intelligence and robotics Photonics Simulation and modeling Computational fluid dynamics

National Critical Technologies[a]	Commerce Emerging Technologies[b]	Defense Critical Technologies[c]
Manufacturing		
Flexible computer-integrated manufacturing	Flexible computer-integrated manufacturing	
Intelligent processing equipment	Artificial intelligence	Machine intelligence and robotics
Micro- and nanofabrication		
Systems management technologies		
Biotechnology and life sciences		
Applied molecular biology	Biotechnology	Biotechnology materials and processes
Medical technology	Medical devices and diagnostics	
Aeronautics and space transportation		
Aeronautics		Air-breathing propulsion
Surface transportation technologies		
Energy and environment		
Energy technologies		
Pollution minimization, remediation and waste management		
No national Critical Technologies counter-part:		High-energy-density materials, Hyper-velocity projectiles, Pulsed power, Signature control, Weapon system environment

Source: *Conversion and Diversification of Defence Technology and Manufacturing*, (London: POST, October 1992), p. 8.

[a] From: Executive Office of the President, Office of Science and Technology Policy, 'Report of the National Critical Technologies Panel,' March 1991.

[b] U.S. Department of Commerce, 'Emerging Technologies: A Survey of Technical and Economic Opportunities,' Spring 1990.

[c] Department of Commerce, 'Emerging Technologies,' Spring 1990.

Figure 3.2
Relationships Among Defence Sectors and the Broader National Industrial Base

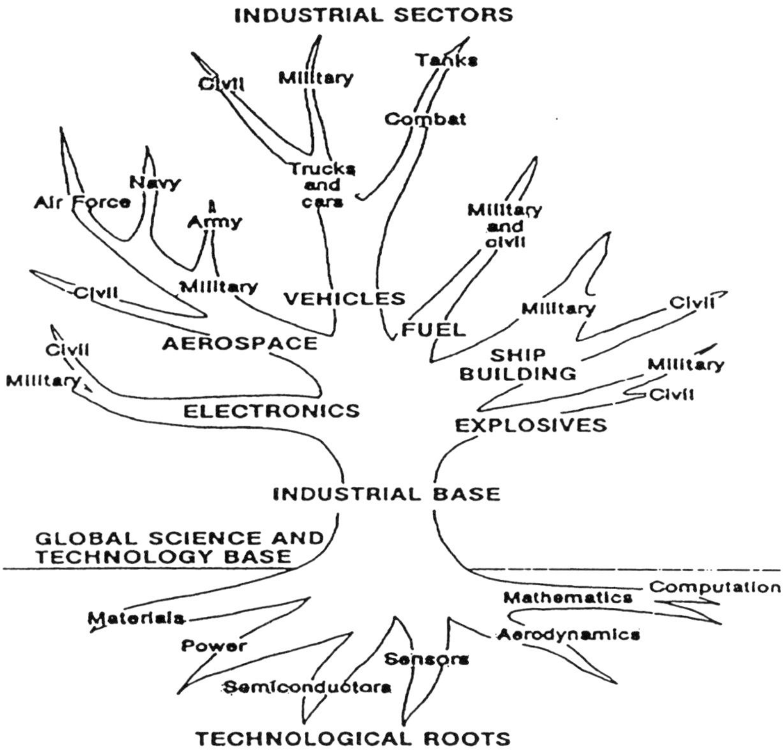

Source: U.S. Congress, Office of Technology Assessment, *Redesigning Defense: Planning the Transition to the Future Defense Industrial Base*, OTA-ISC-500 (Washington, DC: US Government Printing Office, 1991), p. 40.

CONCEPTS OF CONVERSION

Scope

The restructuring of the military, transformation of militarized economies and conversion of military R&D are especially important for the two superpowers. The United States and Russia had to spend vast material and intellectual resources in the most unproductive way during the Cold War. The former could solve many existing social as well as economic problems, if only part of its military potential was converted to more productive civil purposes. In Russia conversion is playing a decisive role within the economic reform program due to the overmilitarization of the entire economy. In other countries, developed and especially developing, a decrease in military expenditures in all fields including R&D would lead to an increase in civilian production and consumption.

At the same time it should be stressed that conversion as such should *not* be considered a global problem. It is the problem of countries with militarized economies due to hot or cold wars. At the same time, however, the many countries that for one reason or another have been spared the destructive developments of militarization are all influenced directly or indirectly by the overmilitarization of some.

Timing of Conversion

The possible motivations for converting military resources, especially R&D, to civil purposes can be divided into short-term national reasons and longer-term international ones. The *short term perspective* is based on changes in national defense policy and procurement planning resulting from the ending of the Cold War. This, in turn, is expected to generate new technological requirements to meet changed force structures and military roles, albeit within tighter military budgets. There is evidence, though, that in some Western countries military R&D budgets are being protected to a certain extent and, insofar as this is so, the pressures for conversion may be mitigated. These changes in national defense policy have implications for the financial viability and employment of arms companies and research establishments and may provide some impetus for diversification, if not conversion. Inga Thorsson, the Swedish expert, emphasized the need for initiating early preparations for conversion because of the long lead-times of up to ten years required for designing and developing competitive products for civilian markets.[9]

The *longer-term perspective* is one of disarmament. Conversion is essential as part of:

- *the process of disarmament and to help reduce opposition (caused by defense job losses) and to create a climate of confidence.* In particular, since Western concerns about previous Warsaw Pact numerical superiority in key nuclear and conventional weapons systems have been addressed in a series of historical disarmament treaties, it is now vital to respond to general concern about the continuing technological arms race. Controlling potentially destabilizing technologies and the research and development of new nuclear and conventional weapons systems should be a major priority for continuing international arms control. No treaty focused on this aim has yet been concluded, although several do contain provisions which may serve to constrain R&D.[10] Conversion of military R&D would be an essential requirement for successful conclusion and implementation of a treaty imposing limitations on force modernization.

- *an altered perception of international security, one covering economic and social as well as military factors.* As long as military supremacy based on technological leadership remains a dominant objective of defense planners it will be hard to give practical effect to a wider notion of security. This is because a continuing arms race will undermine the prospects for radical cuts in military expenditure so that the resources needed to help build economic security will not be made available.
- *broader economic changes, both nationally and globally.* Conversion planning can encourage, and be coordinated with, such changes. By placing greater emphasis on civil applications of dual-use technologies, conversion of military R&D resources might help improve the economic performance of countries like the United States and UK, whose economies have been burdened in the past by excessive military expenditures.[11] Simultaneously, there would be an opportunity to give practical effect to the United Nation's repeated calls for a closer link between disarmament and development by using some of these military R&D resources to address the needs of developing countries.[12]

Level of Solutions

More important even than the timing issue is the level of solutions of the conversion problem. There are three levels: macro, micro, and regional. Some experts consider two levels: local versus national, combining micro and regional. In any case, the local or micro-level is decisive, and the major bulk of the conversion work should be done at this level. At the same time other levels should not be neglected - environmental problems, for example, require deep regional and national involvements. Conversion needs the optimal combination of all three levels for its successful implementation.

It is possible that in the near future a fourth level - the international one - may become more prominent. If international cooperation in this field progresses, treaties and agreements could be signed which would create the legal basis for multilateral steps in addition to unilateral conversion in different countries. It is important, particularly in the case of Russia, that cooperation and conversion are differentiated (as discussed in the section below on international cooperation).

The micro-level concept of conversion can be focused on individual factories and research establishments or the local community. In Russia the factory level is seen at the center of any concept of conversion. The nucleus is the business plan of conversion for each individual enterprise. In the West community conversion, involving public assistance aimed at stimulating the local economy through finding alternative uses for redundant military sites and workers, has been the dominant approach so far. This has been seen as more in line with market-oriented economic policies. Nevertheless, where conversion of military R&D is concerned, the individual research establishment may still be the appropriate focus of attention in the West, too, when it is possible to transfer research and design skills from military to civil purposes.

COMPLICATIONS: PRESENT STATE OF CONVERSION AND FUTURE PROSPECTS

Western Economies

There are many other factors to be taken into account, apart from technological ones, in assessing the possibilities for conversion in different countries. A 1991 review of the conversion literature in the post-Second World War period focused on the prospects for conversion in the six Western countries with major indigenous arms industries: the United States, UK, France, the Federal Republic of Germany (as it was then), Italy and Sweden.[13]

The findings of this study are presented in Tables 3.2 and 3.3, which identify the factors which can either improve or reduce the prospects for successful conversion programs at the macro level and at the level of individual defense firms. The tables enable an initial assessment to be made of the possibilities for conversion from the perspective of Western countries as a whole, individual states, specific industry sectors, defense-dependent regions and individual firms. While some factors listed in Table 3.2 have a greater or lesser relevance for military R&D conversion in particular, taken together they provide the overall context affecting the prospects for any such conversion. Nevertheless, both tables are only intended as a starting point for assessing conversion prospects and do not substitute for a more detailed analysis of the possibilities for conversion at any given level.

This framework, though, does highlight a number of key points. First, conversion planning is often highly dependent on factors beyond the control of individual firms or research establishments, so that they themselves may have little autonomy. This does not prevent diversification, but full-scale conversion is much more problematic. Second, despite the ending of the Cold War, there are many factors which are unfavorable to successful conversion, including the generally poor economic conditions in the early-to-mid 1990s, the continuing arms race and Western governmental emphasis on armaments-based security and deterrence. Third, and most importantly, conversion cannot thrive as a separate entity since the prospects for a viable conversion policy increase as it is integrated into a government's overall policy framework and the interests of the wider public.

PROSPECTS FOR INTERNATIONAL COOPERATION

It is very important to develop mutually beneficial international cooperation in the field of conversion; this is especially important for Eastern Europe and the successor states of the former Soviet Union, since conversion in these countries must take place in the context of dramatic restructuring of their national economies. The first steps are already being made. Considerable work has been undertaken by Russians on the one side and American, German, Italian and French partners on the other. In June 1992 a joint Russian-American declaration on defense conversion was signed which envisages the establishment of a U.S.-Russian Defense Conversion Committee to facilitate conversion through expanded trade and investment.

Naturally, there are many obstacles on this path. Some difficulties spring from systemic differences, and much effort is needed to harmonize the economic bases of the countries involved. Other difficulties in Russia are connected with bureaucracy, inefficiency of management and tremendous corruption. As a result, only $300 million of foreign money was invested in the Russian military-industrial complex at the beginning of 1992. It is clear, though, that conversion in Russia must go on even without any cooperation or financial aid. The military-industrial complex of the former Soviet Union was created independently.

Russia can survive without help and cooperation, but will hardly survive without conversion.

Table **3.2**
Defense Industry Diversification and Conversion: The Context

Prospects for Success	
Improved by	*Reduced by*
(a) Environmental conditions (Western/national perspective)	
Boom *economic conditions* - low unemployment	Recession - high unemployment
Political commitment to greater state intervention in private enterprise economy Government willingness to increase non-defense public expenditure	Political commitment to reduce state intervention in private enterprise economy Government curbs on public expenditure, except defense
Strategic threats reduced - relaxation of *international tensions* - arms control/disarmament agreements signed	Increased strategic threats - heightened international tensions - few or no arms control agreements
Lower absolute levels of military expenditure (inc. R&D) Increased government/public recognition of harmful economic and social consequences of *arms race* Government willingness to give greater emphasis to non-military factors affecting security	Higher absolute levels of military expenditure (inc. R&D) Little recognition by, or influence on, governments/public of harmful economic and social consequences of arms race Government emphasis on armaments-based security and deterrence
(b) Structural conditions (Western/national defense industries or industry sector perspective	
Temporary *historical involvement* by commercial firms in defense work	Permanent defense industry of highly specialized contractors
Reduced *size and extent* of domestic defense industry	Indigenous defense industry maintained or scope enlarged
Weakened *political influence* of military-industrial complex/state management of defense industry	Strengthened political influence of military-industrial complex/state management of defense industry

Prospects for Success	
Improved by	*Reduced by*
Lower regional and industrial *concentration* of defense industry	Higher regional and industrial concentration of defense industry
Many R&D centers in defense industry sector (i.e., competitive *markets*)	Few or no rivals in defense industry sector (i.e., monopolistic or oligopolistic markets)
Less reliance on /potential for *defense exports*	Greater reliance on/potential for defense exports
Increasing reliance on *arms imports* by government	High/increasing level of domestic defense procurement
Higher level of government and industry support for civilian *R&D*	Higher level of government and industry support for military R&D
Less *international equipment collaboration*	More international equipment collaboration (unless international conversion planning is feasible)
Less commercial/state *secrecy* involving the defense industry	More or continuing commercial/state secrecy involving the defense industry
(c) Political conditions (national/regional perspective) (i) State policies	
Foreign policy: détente – agreed link between disarmament and development (in practice)	Aggressive/militaristic posture – no link made between disarmament and development (in practice)
Defense policy: unilateral/multilateral nuclear disarmament or nuclear freeze vigorously pursued Major changes in force structure and procurement requirements Conversion closely linked to a disarmament strategy	Increased military strength and/or limited arms control measures Reinforcement of existing force structure and accompanying weapons platforms/systems Disarmament pursued without regard to conversion planning
Maintaining aggregate demand while reducing military expenditure Low integration of defense industry into *national economic strategy*	No compensation for reduction in military expenditures – strict curbs on government spending High integration of defense industry into national economic strategy

Prospects for Success	
Improved by	*Reduced by*
Strong *regional economic policies* to deal with regions in economic difficulty from whatever cause Closer links between conversion and economic planning for regions	Weak (or no) economic policies to deal with regions in economic difficulty from whatever cause Few or no links between conversion and regional economic planning
Creation of, or extension of existing, public sector, non-defense markets Strong *industrial policies* to deal with industrial sectors in difficulty from whatever cause Closer links between conversion and industrial policy	Reduced public expenditures, especially of non-defense programs Weak (or no) industrial policies to deal with industrial sectors in difficulty from whatever cause Few or no links between conversion and industrial policy
Government role in, and *policy for*, defense industry diversification and conversion Recognition by government, employers and unions of need for advance planning Decentralized approach to conversion planning Legislative and financial support for conversion integrated with other government policies	No government role in, or policy for, defense industry diversification and conversion No recognition by government, employers and unions of need for advance planning Centralized approach to conversion planning No, or separate, legislative and financial support proposed for conversion
(ii) Public pressure	
Strong *public support* for lower defense budgets Strong support for disarmament/arms control	Strong public support for high/higher defense budgets Weak support for disarmament/arms control

Source: Adapted from Peter Southwood, *Disarming Military Industries: Turning an Outbreak of Peace into an Enduring Legacy* (Basingstoke: Macmillan, 1991), Table 14.1.

At the same time, it is extremely important to develop a conceptual model of international cooperation in the field of conversion. At present there are many examples of recommendations given to the Russians, but their value leaves much to be desired. Adelman and Augustine, for example, do not hesitate to propose a number of models for Russian conversion, although they acknowledge at the beginning of their paper that conversion in the United States has a '. . . discouraging history of failure.'[14] They cite an expert report:

> Successful examples of such conversion [of military production in the United States] are difficult to find. Detailed research has not identified a successful product in our economy today which was developed through a military-to-civilian approach . . . As of 1990 there are very few concrete examples of actual conversion.[15]

Table **3.3**
Defense Industry Diversification and Conversion: The Industrial Firm

Prospects for success — *Improved by*	*Reduced by*
Internal conditions (defense company perspective) (i) Relevant to diversification and conversion	
Low *dependency* on domestic defense contracts Low level of dependency on defense exports	High dependency on domestic defense contracts High level of dependency on defense exports
Compatibility between defense and civilian *technologies* - concentration on high technology	Defense technologies not easily adapted to civilian work
Previous *experience* of successes in diversification Proven managerial competence	Past failures in diversification Demonstrable incompetence of management
Future *profits* likely to justify new investment Low *risk* in diversification	Future profits unlikely to justify new investment High risk involved in diversification
Low degree of *existing diversification* - alternative products available	High degree of existing diversification - few more alternative products available

Prospects for success	
Improved by	*Reduced by*
Opportunities for entering new civilian (especially public sector) *markets* available Desire for greater sales than is possible from defense work Innovative ability in commercial markets	Few/no opportunities for entering new civilian (especially public sector) markets Any requirement for increased sales can be met from defense work Little innovative ability/marketing expertise in commercial markets
Availability of *resources* (managerial, technical, capital) for diversification	Existing resources (managerial, technical, capital) heavily tied to defense work
Longer *time span* (5–10 years) for diversification Beginning diversification when defense sales increase	Little time (2–5 years) available for diversification Beginning diversification when defense sales decline
Adaptability to commercial *cost controls*/accounting procedures	Little experience with commercial cost controls/accounting procedures
Acquisition of *personnel* or firms with civil production or marketing skills Labor-intensive production requirements of civilian alternatives	Lack of individuals or firms with civilian production or marketing skills and unwillingness to recruit the same More capital-intensive or less labor-intensive requirements of civilian alternatives
Similar *pay and conditions* of service to that in chosen civilian industries	Higher pay/better conditions of service in defense work
Establishment of *retraining*/reorientation programs	Lack of retraining/reorientation programs
Company *culture*: innovative/entrepreneurial atmosphere Commitment of top management Positive employee attitudes (especially of scientists and engineers) Positive customer attitudes (especially of government) Willingness to accept some product failures	Company culture unfavorable to civilian work No commitment from top management Negative employee attitudes (especially of scientists and engineers) Negative customer attitudes (especially of government) Unwillingness to countenance any product failures

Prospects for success	
Improved by	*Reduced by*
Avoidance of *pitfalls* in defense industry practice, e.g., over-designing, concurrency, poor cost controls, over-optimistic market expectations	Lack of awareness of the need to avoid defense industry malpractices
(ii) Relevant to conversion in particular	
Searches for *alternative products* not limited to those made with existing means of production	Searches for alternative products limited to products which can be made with existing means of production
Business idea (purpose of organization) redefined - compatibility of conversion with corporate strategies	Conservative attitude to redefining the business idea (purpose of organization) - conversion regarded as incompatible with corporate strategies
Commitment to major new *investment*	Unwillingness to countenance any major new investment
Involvement of trade unions (as management may be more interested in defense industry 'pay offs')	Involvement of trade unions prevented or not sought (seen as interference with managerial prerogatives)

Source: Adapted from Peter Southwood, *Disarming Military Industries: Turning an Outbreak of Peace into an Enduring Legacy* (Basingstoke: Macmillan, 1991), Table 14.2.

It is quite questionable to recommend policy on the basis of such experience. What we need is concepts and models of international cooperation in the field of conversion, worked out together by experts of the respective countries and taking into account the differing economic and political contexts in which conversion must take place. Some coordinating bodies for that already exist and should be utilized. The UN system, which has shown great interest in this matter, could be more actively involved in the process.

CONCLUSION

The enormous potential benefit from military R&D conversion is clear enough. Yet for this to happen will require the political will to wind down the global arms race, to radically cut military expenditure in the West (to match that already achieved in the former Soviet Union), and to implement conversion programs to ameliorate the structural problems of economic adjustment. That most of this has not occurred by the mid-1990s testifies to the formidable staying power of militaristic institutions and ideas. It is only through an understanding of these wider obstacles to radical conversion, at a time of unprecedented opportunity for disarmament, that the shortcomings of militaristic strategies of adjustment can be appreciated. Then, if the opportunity is lost, history can truthfully record the reason why.

NOTES

1. See Chapter 6 below.
2. Jan Oberg, 'Is the Conversion Idea to be Converted? Some Sceptical Comments from a Non-Convert,' in *Militarization and Arms Production*, ed. Helena Tuomi and Raimo Vayrynen (Beckenham: Croom Helm, 1983), p. 295.
3. See Council for Science and Society, *UK Military R&D* (Oxford: Oxford University Press, 1986); and Ian Davis, *Military R&D in Europe: Collaboration Without Control?*, Current Decisions Report No. 11 (Oxford: Oxford Research Group, October 1992), p. iv.
4. Parliamentary Office of Science and Technology (POST), *Relationships Between Defence & Civil Science and Technology* (London: POST, May 1991), p. 4.
5. Parliamentary Office of Science and Technology, *Conversion and Diversification of Defence Technology and Manufacturing* (London: POST, October 1992), p. 7.
6. *Western Hypocrisy on Arms Conversion: Helping Military Industries Disarm in the East . . . But Not in the West*, The New Militarism Briefings, No. C/1 (Leeds, UK: Project on Demilitarisation, March 1994), pp. 7–8.
7. POST, *Defence & Civil Science and Technology*, p. 9.
8. POST, *Conversion and Diversification*, p. 4.
9. Inga Thorsson, *In Pursuit of Disarmament: Conversion from Military to Civil Production in Sweden*, 2 vols. (Stockholm: Liber, 1984–85), 1A, p. 301.
10. *Controlling the Development and Spread of Military Technology: Lessons from the Past and Challenges for the 1990s*, ed. Hans Gunter Brauch, et al. (Amsterdam: VU University Press, 1992), p. 7.
11. See, for instance, Robert DeGrasse, Jr., *Military Expansion, Economic Decline: The Impact of Military Spending on U.S. Economic Performance* (New York: M.E. Sharp with Council on Economic Priorities, 1983); and Malcolm Chalmers, *Paying for Defence: Military Spending and British Decline* (London: Pluto Press, 1985).

12. See *The Relationship Between Disarmament and Development*, A/36/356 (New York: United Nations, 1982).
13. Peter Southwood, *Disarming Military Industries: Turning an Outbreak of Peace into an Enduring Legacy* (Basingstoke: Macmillan, 1991).
14. K. Adelman and N. Augustine, 'Defense Conversion: Bulldozing the Management,' *Foreign Affairs* 71 (Spring 1992), pp. 26–47.
15. Adelman and Augustine, 'Defense Conversion,' p. 27.

4

Conversion of Military R&D in the U.S. Context

Adam Yarmolinsky

INTRODUCTION

This chapter presents an overall view of the issues surrounding the conversion of military research and development (R&D). It makes a number of fundamental distinctions: between research and development; between military R&D, which is primarily concentrated at the development end of the spectrum, and civilian R&D; between basic and applied research; and between cost-reducing and cost-enhancing research. It discusses the flow of benefits between military and civilian R&D, in both directions, and it explores the role of the university in military research - a role that may be more highly developed in the United States than in the former Soviet Union.

One way to think about the kind of research and development in which we are interested here is to think about what it is not. To begin with, it is not continuing to do things you have already learned to do. It is not likely to involve devoting more and more resources to a process or purpose to which you have already devoted substantial resources. And it should be distinguishable from activities that are moving you in the same direction in which you have been moving. Instead, research and development should be a source of innovative ideas that serve to advance the state of the economy and improve the lot of the general population.

Throughout history, warfare has probably been the principal stimulus for human invention. From Greek fire and the Roman catapult to jet engines and nuclear implosion, war has provided the impetus for invention. With the dawning of the information age, however, invention as a way to mobilize essentially unlimited resources of knowledge and information has moved to center stage, and the R&D needs of the civilian information economy may either overwhelm the demand for military R&D or - if neglected - leave the United States floundering in the wake of other less war-oriented societies.

Nevertheless, military R&D remains, even after the end of the Cold War, a major force in overall U.S. R&D. Of the total of $157 billion U.S. R&D spending in 1992, $68 billion was spent by the federal government, and of that figure $42 billion was military related.[1] At the same time, we need to point out that most of what appears in the military budget as research and development, or, technically, 'Research, Development, Test and Evaluation,' (RDT&E) goes to the latter stages of the RDT&E process, the engineering of new products or processes for use in the field, and their testing and evaluation for combat effectiveness. This pattern of activity has important implications for the possibilities for conversion of military R&D.

TECHNOLOGY TRANSFER

Paradoxically, the U.S. military establishment spends the bulk of its R&D resources on what the U.S. commercial market for research is often accused of neglecting. The transfer of the results of civilian research expertise into commercially marketable products - that is, technology transfer - is often described as the stepchild of commercial R&D, in contrast to the practice of our commercial rivals in Japan. The military R&D establishment, on the other hand, spends most of its energies on the military equivalent of technology transfer. The work that the military does in the last stages of turning an idea into a weapon system, however, is less likely to be adaptable to turning the same idea into a gadget (or a system) marketable in the civilian economy, than the work that goes into generating the idea in the first place. And the people who work on those latter stages of development are not really engaged in research as we usually think of it.

As a consequence of the different practices in the military and civilian sectors, the opportunities for direct conversion of military R&D may be more limited than might be at first supposed. The question could well be asked, however, whether even if the very considerable human and material resources involved in military R&D are not adapted to civilian research, could they not be diverted to the civilian equivalent of development and testing and evaluation, that is, to technology transfer activities. There is some evidence that some transfer of ideas may be happening. General Motors, for example, has invited 'teams of top researchers and technology transfer officers' from nine major federal laboratories to meet at G.M.'s technical center, after a year of planning in which G.M.'s Advanced Technology Division sent teams out to the federal laboratories, including military laboratories.[2] Military laboratories were given the authority under the Federal Technology Transfer Act of 1986 to sign Cooperative Research and Development Agreements (CRADAs) with industry, and that authority was extended in 1989 to the Department of Energy weapons laboratories operated by private contractors, although bureaucratic difficulties have delayed implementation of CRADAs with contractor-operated laboratories.[3]

The great drawback to this kind of transfer is that military 'technology transfer' is so different from commercial technology transfer that its practitioners are temperamentally unsuited for the commercial world. What they have learned induces a kind of professional deformity, a learned incapacity, that makes it more efficient for large corporations going through conversion from defense business to start over, if necessary, by acquiring new businesses, rather than trying to re-tool their own people. The shift from dealing with one customer to dealing with the marketplace, from price as a secondary to price as a primary consideration, from maximal to minimal consumer involvement in product design, creates a gap too wide to bridge the difference in cultures.

If, however, the proposed conversion is from the military to the civilian governmental market, the transition is much less difficult to effect, and some human resources may be shifted directly. This conversion can free up resources to be applied directly to major public policy problems, in areas where trained human resources are not always readily available. The difficulty here is that non-military governmental R&D is generally not as intensive as the military variety, which inevitably seeks the last increment of technological perfection. Civil government, unlike war, is not, after all, a zero-sum game.

This is not to overlook the enormous contributions made to the civil infrastructure of the United States by the Army Corps of Engineers since its organization in the early 19th century. In fact a logical extension of that contribution was proposed by Senator Sam Nunn, then Chairman of the Senate Armed Services Committee, when he proposed the organization of a Strategic Environmental Research Program drawing on 'the technological resources of the Defense Department, certain elements of the intelligence community, and the national security activities of the Department of Energy.'[4]

BASIC AND APPLIED RESEARCH

At the research end of the spectrum, the human resources devoted to military research are more easily and directly convertible to civilian-oriented research. But here too we need to make several further distinctions. First, there is the distinction between basic and applied research. In principle the results of basic research are universally applicable. They help us to understand the universe. Basic research financed out of the military budget is no more applicable to military purposes than to civilian. Indeed it can be argued that military support for basic research is a net gain for the civilian economy, since it might not be possible to find the same level of support in the non-military budget, especially because it would have to be in addition to existing basic research support from sources like the National Science Foundation and the National Institutes of Health. Just as military exigencies - or at least the general belief in such exigencies - produced the National Defense Education Act, which contributed to the education of American young people, and the National Defense Highway legislation, which created the magnificent network of interstate highways in the United States, so 'military' basic research contributes to the sum of basic research a component that might not be readily replaced if it were taken away. Of course, this is only so if the military research dollars are provided out of a fixed overall military budget, rather than being treated as an add-on to that budget.

The analysis of the benefits of military-funded basic research must be tempered, however, by acknowledgment that the line between basic and applied research is becoming increasingly fuzzy, as dramatized by the current rapid movement of molecular biologists from the basic research laboratories to the counting house. And even basic research is not always interchangeable. Civilian science policy requires decisions, for example, about the allocation of basic research dollars between the natural and the physical sciences, based on some estimate of relative eventual payoff. Military R&D planners also make estimates of eventual payoffs, and their estimates, based on their interests, may differ sharply from those of the civilian basic science planners.

A relatively small portion of the dollars in the research end of military R&D budgets goes for basic research. Most of those dollars go to what is clearly, by any standard, applied research. Here again there are distinctions to be drawn. Some applied military research has positive benefits for the civilian economy. Some does not. For example, software developed for large-scale military applications may not be at all applicable to smaller civilian computing systems.[5] The list of applications of applied military research to civilian purposes, from Landsat photography to microsurgery, is a familiar one. So is the retort that military planners would scarcely be willing to rely on the 'spin-ons' from applied research in the civilian economy to supply the research they need for military purposes.

The fact that there are increasing numbers of examples of spin-on from civilian to military R&D is not surprising. The tempo of the market place encourages more rapid development of invention and more immediate exploitation of the potential of the information explosion. To the extent that military R&D relies on developments in the civilian market, conversion is not an issue. The problems associated with this trend are rather in the area of uncontrollable proliferation to other countries.

Perhaps an equally important distinction is between applied research that tends to reduce the cost of military hardware and of military operations (not always the same thing), and research that tends to increase one or both of those costs. Clearly, the primary objective of military R&D is not to save money, although there is a real danger that the results of R&D can produce weapons so expensive that a military bemused by technological complexity can not afford enough copies to warrant a credible force. Charles Hitch, the economist who first applied the idea of program planning to the military, offers an appropriate parable: the inventors of the bronze spears were so pleased with their product that they substituted bronze for iron, but as a result they could not afford to equip enough of their soldiers with the new spears, and they lost the war.[6]

The tendency to exploit technology for its own sake is a peculiarly American failing. Eugene Fubini, who served as Assistant Secretary of Defense for R&D in the Johnson Administration, called it the American Syndrome: 'If you can do it, do it.'[7] The syndrome manifests itself in civilian R&D as well, in medicine, in automobiles, and in household appliances. And wherever it manifests itself, it distorts the allocation of scarce resources. But in the military R&D context it is particularly insidious because it acts as a multiplier for the inflation of military budgets. A little R&D money can stimulate an enormous production enterprise. In this connection, Les Aspin, when he was Chairman of the House Armed Services Committee, proposed a resource strategy to maintain the industrial base by 'rolling over' technology into new development programs without producing more than prototypes. It is not yet clear whether or how far the Defense Department will pursue this strategy. But development of Aspin's proposal would require a basic change in the culture of the defense industry, in which companies no longer could expect to 'get well' from large development costs through extended production runs.

INTERNATIONAL RAMIFICATIONS

The impact of cost-expanding R&D extends even beyond the borders of the nation where the R&D is conducted. There is a natural impulse to try to share the development costs of a new or improved weapons system with the military forces of allies or even of other governments, so long as they are not currently antagonists of the government that is incurring the initial costs. This impulse to spread R&D costs is a powerful factor in encouraging the spread of new weapons and new weapons technologies to conflict areas around the world, indirectly sparking regional arms races in a vicious cycle. On the other hand, sharing development costs with other governments reduces the net cost for each governmental participant, and therefore presumably lessens the pressure to market the product to third countries.

The situation may be further complicated by unanticipated political developments. When Robert McNamara decided to abandon development of the Skybolt nuclear missile, designed to be delivered from strategic bombers in 1962, because it appeared not to be cost effective for the U.S. Air Force, he did not realize how dependent the RAF was on U.S. development of the missile. Under the U.S.-UK agreement the United States would absorb all development costs, and charge the RAF only for production costs. When the Macmillan government learned that the Skybolt offer was about to be withdrawn, and the only available substitute would be a missile ignominiously labeled 'Hound Dog,' a crisis between the two allies erupted. It was only resolved at a summit meeting where President Kennedy was forced to make concessions on U.S. cooperation with the British nuclear submarine program, going so far that President de Gaulle, in a huff over French exclusion from the rapprochement of the Anglo Saxons, put such obstacles in the path of British inclusion in the European Community that British membership was delayed for several years.[8]

Ever since Skybolt, both the U.S. Defense Department and European Defense Ministries have discussed cooperative research and development programs and indeed negotiated some 35 cooperative agreements. In 1985 Senator Nunn secured passage of an amendment providing the basis for a $200 million program to co-develop and test weapons with NATO allies. The projects developed under the Nunn program ran into difficulties, however, as they neared expensive decisions for production and deployment. For example, in January 1992 the U.S. Army decided to abandon its funding of the development of a so-called 'smart rocket system' because it had been instructed by the Defense Department to pick a single smart munition for development, and it preferred to choose one that was exclusively a United States program. This negative decision served to elicit statements from several NATO allies that future U.S. proposals for collaboration might not be taken very seriously.[9]

Sharing the costs of military R&D designed to produce new weapons can result in international squabbles. By the same token, however, military R&D that reduces the cost of military hardware, or, even more important, the requirements for military manpower, can produce a significant reduction in military budgets and thus work to the advantage of the civilian economy.

Military R&D can benefit the overall military conversion process even more directly when it serves to facilitate an arms control agreement, for example by discovering new or improved verification techniques, or by overcoming environmental objections to weapons destruction programs. Some researchers at the Los Alamos weapons laboratory, for example, are occupied devising environmentally friendly ways to dismantle nuclear weapons, and others are attempting to simulate the work of third-world researchers trying to build nuclear weapons, in order to anticipate proliferation attempts.[10] While some of this research can be and is done within the civilian economy, considerations of access to classified information, as well as of the advantages of inclusion in the military budget, suggest that it is not a likely candidate for conversion.

CLASSIFICATION ISSUES

The issues involved in classified research are probably the most prominent in the congeries of process issues that accompany substantive questions about conversion of military R&D. Classified research is by no means confined to the military establishment. It is commonplace in industrial laboratories as well. De facto classification can even be found in university and non-military government laboratories where researchers are racing for the prize of prior publication. But classification - unwillingness to share results and methods with fellow investigators across institutional and even national boundaries - is fundamentally inconsistent with the spirit of scientific inquiry. That spirit may be inhibited even by the physical juxtaposition of classified military research and unclassified non-military research. Walling off portions of a laboratory compound with designated security areas and all the accompanying security paraphernalia can induce a kind of Bluebeard's Castle syndrome among investigators. The idea seems particularly repugnant on university campuses. Yet one possible consequence of closing some military laboratories, as the military research budget is constricted, may be the transfer of some of the remaining work to universities.

A further concern arises: as military budgets, including research budgets, shrink, research planners in the military establishment may understandably seek to take fuller advantage of unclassified civilian-oriented research in non-military government laboratories and in universities. This tendency goes hand in hand with the increased 'spin-on' to military from civilian R&D referred to above. In order to derive what they see as maximum benefit from this research, they may seek to isolate it from exchanges with foreign investigators, or even to develop informal clearance processes for those associated with the work. While this scenario may sound far-fetched, there have been recent occasions when Defense Department officials have requested universities to exclude foreign graduate students and other foreign visitors from unclassified research sites of special military interest.

Reports from the Energy Department laboratories at Los Alamos and Livermore, both operated under contract with the University of California, suggest that scientists in these laboratories are being constrained by directives that limit attendance at scholarly meetings, particularly with international representatives; filter contacts with the press; and even contain expressions of dissent within the laboratories. As these laboratories reach out for non-military research opportunities, it seems particularly perverse that the academic freedom on which research imagination depends is itself being constrained.[11]

The restrictive ethos of military research is even more likely to impinge on university research because universities as research institutions are increasingly involved in the rest of the world. The academic community has always transcended national boundaries. The international aspect of academic life is quantitatively enhanced by advances in communication and transportation, advances proceeding by orders of magnitude. But there has been a qualitative change as well: the concept of the interactive university has tempered, if it has not replaced, the concept of the university as a protected enclave, an ivory tower. In the information age the university is the engine of economic progress, as the development of energy resources was the engine of economic progress in the previous age. The link between the university laboratory and industrial innovation is immediate. In this climate, preserving the values of free inquiry becomes more difficult, as the economic and political pressures for short-range results are magnified, as are the rewards for beating out the competition. To the extent that the continuing demand for military-oriented research impinges on the academy, it exacerbates all these difficulties.

RESOURCES IN MILITARY R&D

In thinking about conversion, there is a tendency to see military R&D as an area to be scavenged for raw resources - any available dollars, some kinds of people, perhaps some equipment. But there are also some intangible values in certain outstanding examples of military research, that should not be lost sight of in the conversion process. War brings out the heroic instinct, and there are instances of heroic effort in military research, perhaps most notably in the creation of the first atom bomb, whatever one thinks about the U.S. decision to drop it on not one but two civilian targets.[12] The recent history of military research also affords examples of extraordinarily effective collaboration on complex research and development projects that may provide useful analogues for the deployment of R&D resources in the civilian economy.

The first example is a happenstance, but the others were deliberate - and successful. The Sidewinder air-to-air missile, an air-launched heat-seeking missile, has been one of the triumphs of U.S. technology, helping to assure U.S. air supremacy. Yet the Sidewinder was developed at the China Lake research facility by Navy civilian scientists working entirely on their own time, with material and equipment scrounged from the regular work of the laboratory.

The Sidewinder may have been a lucky fluke, but the Lockheed Aircraft Corporation built the U-2 spy plane, and its even more spectacular successor, the SR-71, in a facility deliberately designed and managed along the same lines as the China Lake adventure. Both the U-2 and the SR-71 were the work of a team headed by Lockheed designer Kelly Johnson, at a site popularly known as the Skunk Works, where Johnson and his team operated on their own, free of all the rules developed by the Department of Defense and the Air Force to rationalize research projects. It is not clear how much these successes were due to the genius of Kelly Johnson, and how much they were a result of the freedom he and his team enjoyed to make their own decisions according to their own rules.[13] The lesson to be drawn from the Skunk Works for civilian research, particularly in the new interactive universities, would seem to be that even research closely tied to industrial development (or public service) must resist unilateral regulation by the users, and that researchers must insist on more collegial relationships.

But freedom to do innovative non-military research will be to no avail if trained and qualified researchers are not available. The fact remains that resources devoted to military research cannot usually be employed at the same time in non-military research. Even when resources are freed up by contraction in military budgets, the frictional losses in the process of transition are such that the net gains to non-military research will be less than the net losses to military research – even apart from the fact that so many of the human resources employed in military R&D are actually engaged in development, and, as pointed out above, may be and often are quite unsuited for many kinds of non-military development.

One might argue that the human resource needs of non-military research are supplied by the operation of market forces. This might be so except for the fact that the United States is falling behind in producing technically trained people, and is not able to make up for the shortfall by importing students, and trained researchers from abroad. Impending greater shortages are exacerbated by demographic shifts in the United States and the failure to attract women and minorities to science and engineering in anything like the numbers needed.[14] Impending shortages are temporarily hidden by a surplus of scientists and engineers specializing in military work, and by the macro-economic problems of the world economy.

One way that has been suggested to reduce the impact of these shortages is to focus on dual-use technologies through cooperative ventures between civilian and military laboratories like the Advanced Research Projects Agency, which is generally well regarded as an imaginative research planner. One reason for military laboratories to seek out civilian partners is that Defense Department research money is so widely spread across the country through sub-contractors, that it has no powerful political constituency to protect it in the course of budgetary struggles within the military.[15] The argument is a subtle one: military production contracts are consciously spread in order to generate widespread political support. But military spending for research on generic technologies is so small that it needs to be amalgamated with civilian research in order to be politically visible. Dual-use projects provide a way to make these connections.

In sum, the continuing diversion of human and material resources to military purposes when they might be applied to non-military research and development in the public and private spheres has sufficiently negative effects on both the economy and society that such diversion needs to be justified in very specific terms. Putative benefits to national security, narrowly defined, should be measured against losses to national security more broadly defined, including economic, social and educational losses as is appropriate in the post-cold-war world. This is so despite the current surplus of highly specialized and highly paid engineers and scientists who are being laid off by the U.S. defense industry and in the former Soviet Union, where the situation is far worse. Moving these individuals into productive civilian work will go slowly, keeping time with the continuing slow pace of economic recovery and reconstruction. At the same time it is the next generation of engineers and scientists who need to be attracted to more socially useful careers.

NOTES

1. U.S. Congress, Office of Technology Assessment, *Defense Conversion: Redirecting R&D* (Washington, DC: Government Printing Office, May 1993), p. 7.
2. David Hamilton, 'General Motors to Turn to the National Labs for a Technological Boost,' *Science* 255 (10 January 1992), p. 147.
3. See Christopher Anderson, 'Rocky Road for Federal Research Inc.' *Science* 262 (22 October 1993), pp. 496–98.
4. Sam Nunn, 'Strategic Environmental Research: A New Dimension of National Security,' *Aspen Quarterly* (Autumn 1990), p. 87. See also, U.S. Congress, Office of Technology Assessment, *Defense Conversion: Redirecting R&D*, (May 1993), ch. 3.
5. U.S. Congress, Office of Technology Assessment, *Holding the Edge: Maintaining the Defense Technology Base* (Washington, DC: U.S. Government Printing Office, April 1989), p. 30.
6. Adam Yarmolinsky, *Paradoxes of Power* (Bloomington: Indiana University Press, 1983), pp. 233–34.
7. Yarmolinsky, *Paradoxes of Power*, p. 124.
8. *The Skybolt Affair*, Kennedy School of Government Case Program (16-92-320.0), 1992.
9. *Defense News* (27 January 1992), p 6.
10. Christopher Anderson, 'Weapons Labs in a New World,' *Science* 262 (8 October 1993), p. 170.
11. Eliot Marshall, 'Weapons Labs: After the Cold War,' *Science* 254 (22 November 1991), p. 1101.
12. David Holloway describes the similarly heroic efforts of the Soviet scientists in developing nuclear weapons in *Stalin & the Bomb* (New Haven: Yale University Press, 1994).

13. Adam Yarmolinsky and Gregory D. Foster, *Paradoxes of Power* (Bloomington: Indiana University Press, 1983). See also Ben R. Rich and Leo Janow, *Skunk Works: A Personal Memoir of my Years at Lockheed* (New York: Little, Brown & Co., 1994).
14. Robert E. Fullilove and Philip Uri Treisman, 'Mathematical Achievements Among African-American Undergraduates at the University of California, Berkeley: An Evaluation of the Mathematics Workshop Program,' *Journal of Negro Education* 59:3 (1990), pp. 463–478; Walter E. Massey, 'A Success Story Amid Decades of Disappointment,' *Science* 258 (13 November 1992), pp. 1177–79.
15. This reason was suggested to me by Lewis Branscomb.

5

Transforming the Russian Economy

Ivan Maslennikov

INTRODUCTION

The features and prospects of R&D conversion in Russia are determined mainly by the same factors that underlie the general political economic context of the country's transitional period from a centralized to a market economy. The Russian experience of the past several years convincingly shows that it is impossible to solve conversion problems without economic reforms. What's more, considering the enormous importance the military-industrial complex has for the Russian economy, conversion should be regarded as virtually synonymous with the structural policy of economic reform.

That is why for many Russian experts and specialists there are two conceptions of 'conversion.' In a broad sense conversion is equal to the transformation of the entire economic system and is, in fact, the transfer from the state-controlled and deeply militarized economy to the market economy, a 'mixed' type economy with private and public forms of business activity. In the narrow or traditional interpretation, conversion is usually understood as restructuring the manufacturing sphere of military production. This process, while having its own significance, is at the same time the basis for market demilitarization of the economy (conversion in the broad sense).

Conversion of military R&D plays a special role in this process. It will shake up the whole scientific and technical potential of the country by restructuring afresh all of the research spheres, while laying the scientific and technological basis for future structural changes. Conversion of military R&D today is the structural policy of tomorrow. The intensity of military-industrial conversion – its speed and socio-economic efficiency – depends above all on making 'civilian' the scientific and manufacturing research potential of the military branches. At the same time, the priorities of the state structural policy and the corresponding governmental programs set the general framework conditions for the conversion process, including conversion of military R&D.

Russia inherited the major share of the former USSR's powerful defense complex, which consisted of more than 1100 industrial enterprises and approximately 920 research and experimental institutions with a total work force of more than 9 million people. Eighty per cent of the research and experimental enterprises of the former Soviet Union are located in Russia. In 1992 930 enterprises and about 600 research institutions were involved in the conversion process. They typically suffer from an unstable financial condition because of a lack of financial assets. This makes the conversion program more expensive and delays its realization. Despite the well-known difficulties of the past several years, however, Russian military enterprises and research institutions continue to retain the highest material and staff potential compared to the other branches of the national economy. They are actively involved in most of the essential directions of scientific and technical progress.

It is quite evident today that the economic reforms in Russia, having begun under the banner of a tough monetary policy that at first ruled out state interference and any effective structural and industrial policy, have in fact resulted in a poorly regulated conversion of the defense industry. This is a threat to both the defense capability of the country and the entire industrial and technological potential of Russia, in which the defense branches continue to hold the key role. The single fact that the military enterprises had until recently 60 per cent of all production facilities in Russian industry speaks for itself.

Work on the state's medium-term structural policy has begun a new stage of conversion. From being poorly controlled, the process has become a little bit better organized and manageable. At the same time we are witnessing a transformation in the economic crisis in Russia. Instead of an all-embracing, disastrous fall due primarily to technical causes, such as deficient delivery, shortages and severance of economic ties, it has become a process of structural change. Alongside the failing industries there is some recovery in certain promising, export-oriented industries, which are beginning to gain markets. The task as it stands now is to stimulate areas of potential economic growth, to encourage and intensify the beginning of structural changes, and thus to help the industries in crisis to overcome their problems. This should be the key direction of the government's structural and industrial policy, coupled with a program of measures for the conversion of the defense industry and military R&D. It is those conversion processes that aim at the radical transformation of Russia's scientific and industrial capabilities that bear both the major risks and the greatest potential benefits.

PROSPECTS FOR CONVERSION

Conversion of military R&D is affected by two interconnected and mutually exclusive objectives. The first is the necessity to demilitarize the economy and to make alternative use of the resources made available, while the second and equally obvious objective is the support of Russia's armed forces. The combination of these goals specifies the approach to the conversion of the defense complex's scientific and technical potential. National security concerns are certain to form the basis of such a conversion. Only after the system of national objectives has been determined and the role of the military forces in countering the threat to the country's security has been defined, can one definitively plan the development of arms and weapons and formulate a military and technical policy. Only then does it become feasible to single out that sector of the research and technology base which is to be converted. In the absence of such an analysis all talk about conversion would be nothing but speculation. Nevertheless, making concrete decisions about conversion is quite urgent today, when the overexpansion of military research and development has become apparent and cuts in weapons output make it impossible to use the full capacity of the military and research potential for either defense or 'spin-off' purposes.

In thinking about conversion, three sectors of the military scientific and technological base can be distinguished. The first is that portion of the technology base which is not subject to conversion. It is fairly big in Russia, approximately 20-30 per cent. There are several reasons for this. First, it is impractical at present to completely renounce the use of force or threat to use it. There is still instability in the world, and a dangerous tendency for newly designed weapons to proliferate uncontrollably. Thus national security priorities make the continuing development of military technology necessary. Military R&D provides an efficient response to any possible military threat or technical breakthrough.

Second, the research and development institutions of this core sector are highly specialized and, as a matter of fact, could not be technologically converted in the traditional sense. But under current circumstances these institutions are clearly overexpanded. All this demands delicate and careful converting of these institutions to adapt their research for social needs. One possible direction here would be the use of military scientific and technical resources to create a collective (international) security system. Russian scientists can participate in various international programs for developing weapons or cooperate with collective security systems. Another direction is the development of arms control technologies such as verification procedures for arms limitations and the creation of new international means for inspection of proliferation of weapons of mass destruction. At the same time, the government maintains an ability to improve those weapons needed for its military policy.

Some military R&D resources will have to be assigned to developing and modernizing the technology and equipment used to eliminate discarded weapons by safe and ecological methods. The idea is to make the military R&D institutions work at reverse technologies. It is only natural that the task of the destruction of weapons can best be solved by those who made them, especially in the case of weapons of mass destruction. Basic research, which has always been part of military R&D, would help to implement all these new tasks. Finally, some military R&D is sure to continue in order to serve the needs of military export. Weapons sales for the sake of profits is considered at this time to be necessary, but the sales will be implemented in compliance with international agreements. No delivery of weapons and arms that could aggravate crisis situations will be allowed.

The second sector of Russian military scientific and technical potential is beginning to be used to develop 'spin-off' technologies. This is a case of partial conversion. It isn't so much the substance of the work that is changed, but the attitude to the use of its results, that is, the development of 'spin-off' mechanisms. The most serious obstacle for 'spin-off' is the narrow specialization of much military R&D, which prevents many projects from being used for civilian purposes. In addition, under changing market conditions and the introduction of competition overspecialization will limit the research institutions' economic growth and increase their risks. That is why diversification of military R&D should be the basis of their shift towards the market economy. Diverse research would allow them to find new technological directions without endangering their main strategies. In fact, military design and research centers cannot do without diversification if they are to be competitive. It is essential, however, to find the optimal ratio between diversification and specialization, as overdiversification may lead to an erosion of the core specialties of military research institutions and disperse their forces.

The third sector of military R&D can be fully converted. The main problem here is to determine the potential market. There are different possible approaches. One is to establish big state programs to reallocate resources. Quite logically two questions arise: What is the purpose of the programs, and what are their priorities? One option is that the new production should be directed towards the consumer goods industry. This option looks attractive, because of the seemingly high demand in the current consumer market: in 1992 alone more than one thousand samples of new civilian products and consumer goods have been designed and manufactured by Russian industry. This approach, however, threatens to dissipate science intensiveness and sharply reduce the technological level of the defense industry. Another option thus looks more preferable: not to copy the experience of other countries, but to aim at developing and mastering unique technologies in science-intensive industries. These industries are suitable for the high qualifications of scientists and engineers in the military industrial complex.

It is important to note that the second and the third types of R&D conversion have as their basis those R&D institutions that deal with 'spin-off' technologies and research work. The potential for 'spin-off' has become the main criterion for budget backing. According to the current scientific and technical state policy in the Russian Federation, state support will be given to R&D that develops entirely new types of weapons and military technology; highly science-intensive 'spin-off' technologies; projects connected with economic restructuring; and manufacturing technology for consumer goods.

The 'spin-off' technologies of today are synonymous with the mainstream of the military technological revolution and thus they also serve the security needs of the country. These technologies form the basis for new military means that will have high precision and round-the-clock effectiveness and will be reliable and mobile. Besides purely military benefits, 'spin-off' technologies have some perceptible advantages of a more general character. First of all they provide grounds for considerable cuts in defense expenditure, by reducing personnel needs, introducing new training methods and facilities, etc. But the main saving results from the considerable cuts made possible in the production of conventional arms and weapons and, in some cases, complete suspension of production.

The exceptional significance for Russia of keeping pace in these technologies is that they are for conventional weapons. Military and technical developments are inevitably coming to the point when the unique deterrent functions of strategic nuclear weapons will no longer look unlimited. Creating effective defense systems is a matter of the next ten years. Thus, lagging behind in the sphere of conventional weapons may become dangerous.

The second general advantage to 'spin-off' technologies lies in putting an end to the enclave period of military industry development. The industry's effectiveness is now directly dependent on the overall economic, scientific and technical development. It is true that in Russia the defense industry, particularly some of its branches, possesses the most advanced technologies. But the industry will soon be behind those of other countries unless speedy general recovery begins in the near future. Moreover, the worldwide tendency for leveling of military and state branches is partly present in Russia, too, stimulating both industrial integration (increased closeness of branches and enterprises) and diversification of industrial and research structures in the defense sector.

Developing the 'spin-off' process requires state measures for spreading R&D developed in the military sector to the civil sector. These measures require the following:

- legal basis for technology transfer to the civilian sector;
- institutional structures responsible for the process.

It is necessary to pass new laws to regulate the process of technology transfer from military programs on a commercial basis. Legislation should provide for the establishment of joint research centers for developing technologies for both military and civil sectors. Research institutions and defense ministry centers of the Russian Federation should set up offices which would look for 'spin-off' technologies and promote them.

The economic process of redirecting military research institutions to produce new goods is very complicated, as the approaches facing military R&D in military and civilian spheres differ greatly. When developing and designing new types of weapons military R&D was required to provide given tactical and technical features; costs were of secondary importance. The former economic system in fact guaranteed this approach. Military research institutions enjoyed special privileges practically without any competition, and they were not concerned with survival under market conditions.

One of the special features of the economy of the defense complex is that research and development expenditures represent the biggest and ever-increasing cost of producing weapons, considerably exceeding the actual production costs. The sharp increase of military R&D expenditure has a number of causes. First is the increased technological complexity of new weapons systems, especially heavy weapons. According to some sources, it now takes from 10 to 20 years to develop and deploy a new weapons system. And the labor intensity of developing new weapons is also growing. Second, the growing share of unproductive expenses contributes to the increasing cost of military R&D. These expenses are linked with duplicate developments and unfinished and unsuccessful developments, as well as 'technical uncertainty' of a project. Third, the tiered system of R&D performers, which consists of the leading contractor and main major subcontractors, results in accumulating costs and profits. It affects the price of both the experimental and production samples. Fourth, there is rising inflation, with increased cost of research equipment and growing expenditure on staff salaries.

Because of the above-mentioned factors, the firms and offices engaged in military industrial complex have traditionally enjoyed some distinctive features. They are:

- the opportunity to finance R&D projects that don't give rapid commercial benefit;
- practically guaranteed payment of signed contracts without the need to market the new products;
- higher salaries, on average, for scientists, engineers and managers;
- access to information and new technologies belonging to the federal government;
- the chance to ameliorate a number of social problems of towns and regions at the expense of the Defense Ministry.

These peculiarities, which were fully and vividly manifest under the centralized planned economy, do not encourage cuts in military expenditure for R&D or the conversion of this sphere. Still other circumstances strengthen the inertia in the current system of funding R&D in Russia. In the first place, there are wide-spread fears that conversion measures will be carried out speedily and incompetently. The military scientific and technical potential remains one of the few factors that still make Russia a world power. At a time when scientific and technical progress influences both the economic situation and national security, any threat to lose this potential is seen as a blow to national dignity.

In the second place, an important feature of the scientific and technical potential of the defense complex is its scattered location, with peak concentration in remote and formerly closed towns and territories. It is not only those towns that the public has never heard of because one couldn't find them on any map (e.g., Arsamas-16, Cheliabinsk-70, Krasnoyarsk-35), but also some satellite-towns of Moscow - Zelenograd, Kaliningrad, Zukovsky and others. Whereas the acute social consequences of conversion in Moscow and St. Petersburg could be softened by a wide range of opportunities for the freed-up military-economic resources, the situation is quite different in those towns where there are no possibilities for using scientific, technical or human potential outside the defense research institutions. In those places curbing military research without special financing and employment schemes is fraught with social consequences.

In the third place, it is necessary to consider the economic and psychological aspects of employment in the defense complex. For many years it was the most prestigious occupation in the Soviet Union. The scientists engaged in military research and developments felt themselves part of a great cause to save the Motherland and mankind from a nuclear war. And they - nuclear-physicists, missile-builders, space engineers and other experts - accomplished their task. Though initially lagging behind, they managed to establish military and strategic parity with the USA, and they had every right to be proud of it. These people feel very uncertain when it comes to developing equipment for the food industry or for consumer goods. Nevertheless, under emerging market conditions and new demands, 'scientific-war' institutions have to transfer intellectual activity to other vital civilian tasks. They are obliged to carry out an economic assessment of their products, consider their ability to compete, bring down their costs, and increase profits. What's more, growing competition forces the institutions under conversion to provide civilian consumers with scientific products of higher quality at the same cost and without increasing their prices.

NEW DIRECTIONS IN CONVERSION POLICY

When discussing the conversion of military R&D it is essential to remember that the reform process should be compatible with its innate character; to use inappropriate methods is to ruin the process. It naturally follows that the effectiveness of R&D conversion depends directly on fitting the process into the restructuring of science and the economy in the Russian Federation.

In line with the general concept of economic reforms in Russia, it is planned to demilitarize both the military industrial sector and military R&D spheres by developing alternative forms of property and enterprise, cutting state subsidies and shutting down ineffective production and institutions by lifting restrictions on foreign competition. The main conditions for enterprises to qualify for financial, material, and technical assistance from the state or support from domestic or foreign investors are structural and legal changes in ownership and commercialization in order to perform efficiently under market conditions. To break defense complex capacities out from the sectoral system of management, independent corporations and concerns will be set up. There will be state assistance for raising funds from sources other than the state - preferably from foreign investors - to carry out conversion. The most promising direction is likely to be operational development, with foreign participation, of scientific and technical defense achievements oriented towards civilian use for both home and world markets. This direction offers very good opportunities to the aviation, space, missiles, and shipbuilding industries as well as the communications industry.

The existing specializations of defense enterprises and institutions cover a broad range of scientific and technical directions. Though financial restrictions do not allow the state to support all the directions, the state will assist those that are able to do the following:

- maintain the core of accumulated scientific and technical potential in the defense complex;
- continue the work in directions which either yield world-caliber results or promise commercial success;
- make maximum use of 'spin-off' technologies;
- join in the international system of division of labor in the scientific and technical sphere;
- create new jobs; manufacture import substitute or export products; or attract foreign investment, thus diminishing the economic crisis and encouraging recovery.

Preliminary assessment shows that the following scientific and technical fields meet the above requirements:

- basic research, which is of prime importance for saving the scientific potential of the country;
- chips, opto-electronics, and component parts for electronics;
- new materials, new processing and manufacturing technologies for production of parts and assemblies, including flexible automated production technologies;
- software system and computer-data programming facilities.

At the same time state subsidies will be reduced for those areas of technology that are 10-15 or more years behind the rest of the world. Among these are some electronic and computer technologies. For these fields it is worthwhile to borrow technologies from abroad, diverting their resources to other fields of research while leaving the necessary minimum to absorb foreign experience.

These priorities must be decisive in allocating state subsidies among the scientific and technical institutions of the defense complex. Analysis shows that not more than 40 institutions constitute the core of the scientific and technical defense complex potential. Basic research centers will be the first to receive state support. Among other changes they will be withdrawn from the defense complex and given the status of state scientific centers. Conditions will be set for other institutions to encourage business and innovations on a commercial basis.

The performance by LUCH, a scientific and industrial concern of the Ministry of Atomic Energy of Russia, can serve as one example. During the first two years of conversion the concern set up production of high temperature equipment made of silicon carbide, which is manufactured by just two other companies in the world, Norton (US) and Toshiba (Japan). The output capacity fully meets the demand of Russian industry and in 1992 export of the product to European countries began.

EXAMPLES OF CONVERSION OF R&D

As a result of 'scientific-war' conversion the civilian research share in the defense complex increased from 39 per cent in 1989 to 49 per cent by 1992. Thirteen special conversion programs are under way, among them a program of development of civil aviation under which a new generation of aircraft of the type of IL-96-30, TU-204, IL-114, TU-334, AN-74, AN-38, etc. will be completed and production will commence. The technical characteristics of these aircraft correspond to world standards for the corresponding type of aircraft. One should note that R&D is 62 per cent of the total expenses for the program.

The Russian fleet revival is aimed at expanding the production of civil transport facilities by equipping vessels with modern radar, navigation, power and electrical systems. There were also plans to build 158 sea vessels of total deadweight 1,404,000 tons, 419 fishing vessels of total capacity of 300,000 HP, about 165 river and river-marine vessels of total cargo of 650,000 tons, and others by the year 1995. Before implementing this program the volume of military shipbuilding was to be cut in half. The freed resources will be used for equipping civilian ships, thus increasing their competitiveness. R&D is provided to develop the equipment for ships and devices and technologies for shipbuilding that will meet international requirements.

The program for communication technologies and information science provides for the development of new satellite technology, radio-link communication and television, building new generation computers, and setting up territorial information centers to offer information services to banking and taxation systems, to law enforcement offices and to other commercial businesses. This program gives special attention to measures to bridge the gap in the telephone system. Russia rates 27th in the world in number of telephones per 100 citizens, and the level of external communication services is 10-15 times lower than in developed countries. The program stipulates development of satellite communication and television, setting up production of modern ATS and telephone units and a new generation of radiotechnical and television equipment.

A number of programs make provisions for setting up and developing production of equipment for the processing branches of the agricultural and industrial complexes, light industry, and trading facilities.

In January 1993 the program called 'Ecology Conversion' was initiated. The program provides for the development and serial production of the following products:

- new technologies and equipment for purification of exhaust gases and sewage water; processing, decontamination, and disposal of solid waste;
- measurement and information systems for detection and monitoring of environmental pollution;
- new ecological, energy, and resource-saving technologies.

The program also provides for the development and application of methods for repairing environmental damage resulting from economic and military activities, setting up the material and technical base for an ecological control system to monitor the marine environment within the territorial waters of Russia, and restructuring the economies of the regions suffering from environmental crises.

The aim of the program for the fuel and energy complex is to set up the production of modern gas and oil field equipment, mining equipment, and power-producing equipment using available military industry capacity. It is expected to prevent the fall in oil production and will help to increase it in future. Thus, in 1993 defense industry enterprises manufactured a mobile complex of oil formation fracturing equipment developed by the Heat Engineering Institute (Moscow).

Implementation of the conversion program of enterprises under the Ministry of Atomic Energy of Russia promises to have considerable impact. Its main directions are: medical equipment; communication systems with fiber optical components; manufacturing engineering for the agricultural and industrial complex; and microelectronics, computer, and automation equipment; as well as a set of measures to rehabilitate nuclear contaminated areas.

The above mentioned state programs establish a framework for military R&D conversion. By setting conversion priorities they put the process in motion. Thus one of the main drawbacks in the early conversion activities, the almost complete disregard of the military R&D sphere, is being overcome.

CONVERSION OF BASIC SCIENCE FACILITIES

Radical structural changes in science in the Russian Federation have occurred since the end of 1992. They are aimed at a new scheme of management based on a combination of big research projects financed by the state or big business and a large number of separate technical innovation decisions carried out by enterprises and research institutions.

Following an inventory of all research institutions a limited network of national scientific and technical centers and non-profit research institutions has been set up. They will participate in the state program 'Technopolis' and carry out basic and applied studies. The government encourages the development of 'in-house' science, setting up a network of technopolises, technoparks, and small companies to form an infrastructure of research-innovation activity. Twenty-two research centers will be established. They will retain and develop the scientific and technical potential of the Russian Federation in the fields of nuclear energy (including nuclear waste disposal), aviation, meteorology, optics, laser engineering, shipbuilding, hydrography and hydrophysics, power electrical engineering, automotive engineering and production, ammunition, fuel, explosives, new materials including high-frequency, high-energy physical chemistry, virology, and others. The total number of employees of those centers was about 100,000 by the first quarter of 1993.

The Central Aerohydrodynamic Institute (CADI) is an example of the 'ideology' of such centers. CADI was founded in 1918. In 1930 the government decided to move the institute from Moscow, and the town of Zukovsky appeared. Its appearance marked the start of all present-day aviation science in Russia. The Aviation-Research Institute, the Myasitchev design office, the Aviation Equipment Institute, branches of all the aviation design offices, and faculty from Moscow's physico-technical institute and Moscow's aviation institute department are situated there. There are 65,000 inhabitants, 75 per cent of whom are employed in the town's aerospace complex.

The experience of conversion in Russia leads to the conclusion that science-intensive technologies should be used for their original purpose if possible. That is why CADI adopted a policy of manufacturing civil aircraft and transport aerospace systems. CADI's unique research base is now a federal property and gets budget financing. Some experience in diversifying the Institute's developments to other branches of industry has already been accumulated. There are more than 60 of them now. In order to expand its sources of financing, the Institute has proposed to the government that the part of the aerospace complex which is not the unique national property be privatized.

Further development of CADI will lead to the creation of a technopark and science-city, the major purpose of which will be personnel education and training. There are plans to open a regional University, which would give the citizens of Zukovsky and other nearby towns a chance to get the education necessary to be employed in the aerospace industry. Aerospace enterprises form another section of the technopark. Many of them are situated outside Zukovsky-town. Still another section is a system of private affiliated companies that manufacture industrial and consumer goods. In order to transform the city into an international technopark it is planned to create a database of high technologies available for use by any interested company.

CONCLUSION

The study of the developing process of military R&D conversion allows us to conclude that development and production of weapons for the Russian Armed Forces will be carried out by a limited number of highly specialized defense institutions and big, diversified private corporations, which will get military orders on a tender basis. The majority of the defense complex institutions will be involved in 'spin-off' from the military to the civilian sector to contribute to the high technological level of Russian machine-building, to win and maintain the leadership in certain spheres or scientific and technical progress, and to establish Russia's position on the world market. It is hardly necessary to point out that careful regulation of the conversion process and implementation of the approaches described above will greatly minimize the risk of this process for the world community.

6

The Dashed Hopes for Conversion of U.S. Military R&D

John Tepper Marlin[1]

INTRODUCTION

The end of the Cold War presented a tremendous opportunity for rededication of much U.S. Government spending on military R&D. Around $35 billion in fiscal 1995, the military R&D budget is a significant public investment, supporting thousands of projects in hundreds of facilities and leveraging significant private-sector resources. If rededicated, these resources might provide solutions both to America's economic stagnation - four out of five Americans are experiencing relative economic decline - and its environmental problems. But rededication would mean shifting a substantial portion of military R&D resources to other pressing national and global needs.

A president was elected in 1992 who supported such a rededication. President Clinton came into office with a plan to spend nearly $20 billion on a Defense Reinvestment and Conversion Program during fiscal years 1994 through 1997. Much of the spending was for technology-related programs that would civilianize military R&D. Even discounting some relabeling of military spending, it was the most ambitious conversion program the United States has ever had, with something more than $8 billion of genuine conversion money.[2] The tragedy for the world, not just the nation, is that entrenched institutions and the political realities of 1993-1995 did not permit this program to happen. Congressional action has cut budgets and programs for conversion, and as of October 1995 the outlook for conversion of military R&D is poor.[3]

The failure of rededication efforts is a serious loss for the nation because military R&D still accounts for more than half of U.S.

Government R&D spending and about one-fourth of all U.S. R&D spending. Military R&D has been cut somewhat, but freed scientific and engineering resources are not being rechanneled to any national purpose. The number of scientists and engineers in defense-related work could drop by 40 per cent in the 1990s; almost 100,000 R&D jobs had been lost in civilian defense-related R&D alone by 1992.[4]

In this chapter, we ask and attempt to answer five questions:

- How did we lose the peace dividend in the R&D arena?
- Why is conversion of military R&D inherently so difficult?
- What private and governmental mechanisms and theories were advanced to facilitate the process of converting military R&D?
- Why did they fail to capture the national imagination?
- What mechanisms exist or might be created, within countries or internationally, that could be used to promote conversion of military R&D?

The case of the national laboratories is treated separately by Ray Kidder in chapter 11 of this volume.

WHY IS CONVERSION OF MILITARY R&D SO DIFFICULT?

Characteristics of Military R&D Personnel

Obstacles to conversion are inherent in the composition of the R&D budget. In the United States the bulk of military R&D, historically more than 60 per cent, is performed by industry. Another third is performed at federal labs, and the rest is performed by universities and other non-profit institutions.

'Military R&D' itself is something of a misnomer because only 3 per cent of military R&D is 'R,' basic research; another 10 per cent is applied research, and 86 per cent is 'D,' development; the remaining 1 per cent is dedicated to facilities construction and maintenance. By contrast, for non-military R&D, 38 per cent is basic research, 31 per cent is applied research, and 26 per cent is development.[5] The high percentage of R&D spending dedicated to development means that the work is highly specialized and technical in a way that is specific to the project. Individual workers and managers who have been in this kind of work for some time become decreasingly able to change the purpose and routines of their daily work. While basic research is also specialized, its purposes and applicability are more general, and scientists in such work have little difficulty changing projects. For one thing, basic researchers are in demand as teachers.

A few years ago, the distinction between basic researchers and development workers was not often made. The major shortcomings attributed to military-industry personnel are ignorance of cost and ignorance of civilian markets. Jacques Gansler and others, following the lead of Seymour Melman, argued that military scientists and engineers across the board become spoiled within a few years of entering defense work by the cost-plus and quality-first environment and are therefore difficult to retrain for civilian work.[6] As a consequence, companies like Boeing have in the past kept their military and civilian divisions separate, and movement from military to civilian divisions has been rare.[7]

The Melman-Gansler position, and the conventional wisdom it helped inspire, discouraged conversion efforts: if military personnel are spoiled for civilian enterprises, why bother trying to retrain them? This conventional wisdom has been challenged by Harvard Business School professor Joshua Lerner, who concludes from historical data that defense companies have much to gain from spillover of military technology and that civilian activities benefit greatly from rotation of military scientists and engineers into civilian work.[8] Lerner shows, however, that transfers from military to civilian work are more successful among scientists doing basic research than among engineers working on development. They are also more successful in companies with a relatively low percentage of their work in military contracts, and in companies that deal in dual-use technologies (i.e., products or systems capable of serving either military or civilian purposes). So the closer to final production a military employee is, the harder it is for him to make a successful transfer to civilian work. However, it may also take a significant passage of time for scientists and engineers to become wedded to military R&D work. Lerner shows that they move in both directions in large numbers - of defense scientists and engineers, fully one-quarter crossed the boundary to or from civilian work in a four-year period.

A different explanation of Lerner's research is that conflicting theories and findings about the mobility of scientists and engineers may be explained by the management of their work rather than in deficiencies in their knowledge. The managerial climate in the defense sector differs greatly from the commercial sector, so that working habits in military projects are a poor guide for making commercial choices. This difference may be most important for those scientists and engineers in management positions; engineers at lower levels should have an easier time adjusting. Hence the transitional difficulties reported in interviews may apply only to project leaders. This suggests that the difficulties of converting from military to civilian production may not be as great as perceived by high-level managers, and that conversion of R&D requires breaking up management structures and moving technical people, rather than trying to move entire teams.[9]

We can conclude that recent thinking and observation give conversion programs in science a higher probability of success than conventional wisdom assumes. At the research end of the R&D spectrum the success rate of conversion is likely to be high and doesn't require government training assistance. The need for federal training money for civilianizing scientists and engineers is greatest for development personnel in military production, where the success rate is likely to be lowest.

Regional Economic and Political Factors

Military R&D in the United States is highly concentrated in certain locations. National laboratories and university research centers are concentrated in a few dozen locations. High-technology companies follow these centers and cluster around them. Thus some users of 'multipliers' based on input-output tables have argued that federal spending on military R&D has a higher multiplier than equivalent civilian spending.

Areas where military R&D is based are especially dependent on military spending because of several characteristics of military R&D:

- Much of it is highly specialized, lacking civilian counterparts. This limits mobility of personnel.
- Much of it is secret, so that R&D facilities are controlled and the access of scientists and engineers to others in their field, to the press, and to scientific publications is restricted. This also limits the mobility of personnel.
- For secrecy reasons, some military-research facilities are physically isolated from the rest of the world - Los Alamos being an obvious example.

Because of this occupational and sometimes geographic isolation, dismissed workers and local suppliers may have few alternatives when defense contracts are cut. Those in the labs and universities will be most affected by the state of the academic market for scientists at the time they are looking for work. Those at the development end, more likely to be in a corporate environment, are even more vulnerable. The process of getting accustomed to civilian labor and technology markets takes time.

National conversion assistance can speed up the process and thereby reduce the wastefulness of the transition.

Military-Industrial Lobbies

Any economic approach must take into account the political process, in which the leaders of R&D institutions use their ample political clout to protect their projects. The lawyer-heavy and science-short Congress has historically lacked the expertise to evaluate projects; only four members of the 1993–94 Congress had scientific credentials, and the leader of this group, Rep. George Brown, former Chairman of the House Science and Technology Committee, lost much of his influence with the election of a Republican majority in the 1995-96 Congress.

The assets dedicated to military R&D represent billions spent every year on top of the trillions invested since World War II. The resulting patents, facilities, and - most importantly - the skills base offer tremendous potential for improving the quality of life for Americans by addressing long-neglected and newly emerging national missions. But the debate in 1995 on Capitol Hill was for the most part between two kinds of hawks:

1. The deficit hawks, who would reduce federal support for anything, including both military and civilian R&D, in order to reduce the annual federal deficit.
2. Traditional military hawks, who ignore the end of the Soviet threat and seek to return to the high military funding levels of the Reagan era, especially by recreating the Strategic Defense Initiative (SDI or 'Star Wars') under the new name, the Ballistic Missile Defense Initiative.

Behind the public image of the debate over ideology, military lobbyists behind the scenes continue to promote the interests of different actors. Military R&D funding has a range of lobbyists. Most of the military R&D funding that flows through the Defense Department is targeted to specific sectors. In 1990, for example, more than 51 per cent of all federal R&D funds went to aerospace, most of it for military research. The $17 billion federal investment accounted for 72 per cent of R&D in the sector. Electrical machinery and communications received $6.1 billion, most of it military, accounting for 38 per cent of all R&D funding in the sector.[10]

Because of the high concentration of military R&D spending in a few industries and locations, the affected companies and communities are highly aware of their vulnerability and devote substantial efforts to ensuring that their voices are heard in their state governments and in the national government. The regional and institutional clout of the national labs and university research centers depends on the prestige of the experts who run them, and their access through their academic origins to other well-placed experts in universities and governments. Money also helps. In 1990, Common Cause reported that defense-industry PACs and officers gave $7.9 million to support their candidates.[11] Most of the money is tied to promoting specific systems in which corporations have an interest.

THE CLINTON ADMINISTRATION CONVERSION MODELS FOR R&D

The Clinton administration arrived with the intent of increasing federal involvement in R&D based on the argument by some economists that free market incentives alone 'cannot sustain adequate spending on R&D.'[12] The administration considered a large menu of strategies, many promoted by their advocates in Congress - including basic research, precompetitive consortia, generic technologies, scientific infrastructure, industrial extension, high-risk path-breaking technologies, strategic competitive technologies, and large demonstration projects.

During the debates over these government programs, three main views emerged: (1) the public goods view, (2) the institutional view, and (3) the behaviorist view.

The Public Goods View and 'Big Science'

The market view, based on classical economics, holds that the free market is the optimal mechanism for capitalizing on innovative ideas. Any scientific discovery with a potential market value will be seized on and developed by market actors.

Government has a job to do, however, because of the 'free rider' problem that results from the public goods character of knowledge. Private firms won't invest much in basic research because they cannot appropriate the benefits, so government must do it. In particular, government's job is to prime the pump of innovation by spending large amounts of money to generate scientific discoveries by funding 'big science' projects such as sending a man to the moon. Then private firms will latch on to the discoveries unleashed by federal projects and will disseminate them, with pens that write upside down and new kinds of food packaged to be eaten in space. Advocates of this approach believe that almost any 'big science' project will generate or support innovation.

U.S. lobbyists have often made use of these arguments to justify large federal projects. Advocates of the Cold War arms buildup have maintained that 'technology developed under defense auspices has influenced commercial practice' by a process called 'spinoff,' a kind of trickle-down or trickle-out theory of benefits flowing from military expenditure to civilian use.[13] Now, with the downsizing of military R&D, advocates of large-scale conversion projects argue that equivalent investment is needed directly in civilian R&D.

Clinton administration members and allies have proposed large domestic demonstration projects like the creation of magnetic levitating (Maglev) trains. The National Research and Education Network, otherwise known as the 'data superhighway,' may be the most popular White House proposal.[14] Championed by Vice President Gore, its appeal lies not only in its status as an impressive demonstration project, but also in its potential impact on scientific infrastructure, a more recent objective.

Having funded scientific discoveries, whether through academic research or the creation of sophisticated gadgets, the federal government in this approach speeds circulation of those ideas. In the free market of technological innovation, the other requirement for government is that it get out of the way, eliminating any statutes, regulations, procurement specifications, or standards that might limit incentives to capitalize on new discoveries.

The Institutional View

Opponents of the first view stress the vulnerable position of institutions in the world economy.[15] This vulnerability is often used by leaders of R&D organizations - company executives, research institute directors, federal lab administrators - to account for obvious failures of the market to capitalize on all new discoveries. Even in the R&D-rich defense industry, 'the unaided and unplanned diffusion of defense technology to commercial companies is simply too slow, too inefficient, and too narrowly restricted to a few defense-related industries to constitute an effective strategy for competitiveness.'[16]

Institutionalists stress each institution's philosophy as the key to its receptiveness to new ideas. They argue that the market encourages groups to ignore innovation because competition forces institutions to hoard their technical knowledge, mistrust discoveries from other sources, wait for markets to develop before developing new products, and focus on short-term prospects rather than research requiring more time and higher risk.[17] They conclude that rather than relying just on large federal projects, explicitly collaborative groups like R&D consortia, encouraged by the government, are needed for technology transfer. 'A technology of collaboration, which balances the ongoing competitive tensions among members, is essential for successfully traversing the episodic shifts of the R&D process.'[18]

The government's role in this process is not so much its capacity to marshal resources to sponsor large scientific projects but rather its ability to facilitate collaboration and to act impartially. Like a mother hen tending to its chicks, government is seen as tailoring its innovation aid to the needs of each group, bringing new technology through such mechanisms as development centers, 'buy-in' public-private partnerships, and extension programs, or uniting hostile firms in fields where shared research might prove beneficial.

The Behaviorist View

The behaviorists are even more pessimistic than the institutionalists and believe that technology transfer depends on an even smaller unit, individuals and groups within institutions. They assert that transfer fails even where competitive barriers have been surmounted and suggest that working behavior, not technical merit, determines which new technologies an institution accepts.

A connection can be drawn between behaviorist thinking and the previously noted view that conversion of military R&D 'requires breaking up management structures and moving technical people, rather than trying to move entire teams.' The obstacle in both cases is the manager who is used to solving technical problems in accustomed ways. The lower-level workers can be transferred without too much pain, but for the managers, 'inserting new technologies requires not only a change in their working behavior, but . . . in cognitive processes as well.'[19] A flow of ideas between people and groups must accompany new innovations: 'We view technology transfer as a people-to-people activity, a contact sport.'[20]

To guide technology transfer under these conditions, behaviorists offer two principles:

- Communication networks must interconnect all parts of the institution, including front-line employees not normally involved in technology. 'Technology . . . seems to transcend traditional functional relationships. A 'peer-to-peer' networking system is a communication tool that encourages the cross-organizational learning [that is] required.'[21]

- People must meet face-to-face and participate in group meetings to facilitate a 'cross-pollination' of ideas. The less personal and participative transfer becomes, the less effective it is likely to be.[22]

Given these principles, the proper role of government in technology transfer, if any, is modest or at best unclear. Advocates encourage incorporating workers into the product development process and they champion private firms that have been able to restructure themselves to capitalize on innovation. But virtue in this case should presumably be its own reward. Companies that follow the principles and do the right thing should prosper. The federal government might well disseminate the principles, but few have seriously proposed that federal agencies should go further and attempt to impose them on, or even facilitate their introduction in, private institutions.[23]

Policy Implications

While the behaviorists serve primarily as a warning that technology transfer is harder than it seems, the other two viewpoints have actively been promoting their projects in Washington. Advocates of 'big science' in the Clinton administration promote data superhighways and bullet trains. Those in Congress favor revival of Star Wars. Skeptics note that national governments have failed to pick economic winners in such recent efforts as synthetic fuels, supersonic transport, the Clinch River breeder reactor and the space shuttle.[24] They note that even sound R&D possibilities are often twisted by politics, as constituencies form to increase and unnecessarily prolong federal funding. Opponents call the 'big science' projects a 'technology pork barrel.'[25]

The institutionally oriented advocates shy away from large demonstration projects, preferring programs with public-private cost-sharing, a diversified list of potential products, and independent evaluations. For large companies, they propose precommercial development consortia to control risk and expense; for smaller firms, they propose industry extension programs that bring technical advances to isolated, low-capacity operations. Of course, government would coordinate these programs, prompting some of the same criticisms about Washington's ability to pick the right technologies. Having government select target areas is risky. Harvard economist Robert Lawrence said: 'It doesn't mean there's nothing government can do, but you get precise at your peril.'[26] Meanwhile, as the United States moves after the Japanese model of government support of precommercial development, Japan is having second thoughts and is moving back toward basic research.

MECHANISMS FOR CONVERTING MILITARY R&D

The problem of converting research operations from one task to another is not unique to military conversion. For example, in a well-known example of 'conversion,' when Jonas Salk's vaccine virtually ended the scourge of polio the March of Dimes shifted its attention to birth defects and other childhood illnesses that are respiratory in origin, and did so effectively.

In this section we look at mechanisms for converting military R&D that are in place in the U.S. Government, in universities and private companies. We can begin by noting that the conversion process would benefit enormously from better information on military R&D. Such information is needed to: (1) give civilian companies adequate access to government-subsidized technologies for possible use in civilian markets; (2) identify options for government policy; and (3) identify promising market niches. Information on successful conversion efforts can provide examples for others to emulate. The Economic Conversion Information Exchange in the Department of Commerce, easily accessible through Internet menus, is therefore a welcome addition to available information resources.[27] Another good source is the Bonn International Center for Conversion (BICC).[28]

National Initiatives for Conversion of R&D

Proponents of market solutions argue that national governmental action is not required to address the problems of conversion, that markets for companies, facilities and labor will absorb these components of the military entities most efficiently. Certainly, the transition is easier in a market economy than it is in Russia and the other newly independent states, both because of the rigidity of the labor market left over from Communist days and because hyperinflation wiped out domestic capital resources. But the market also has its limitations. As in Russia, the U.S. government must play a role in the orderly reduction of military spending because – as we saw in the 1992 presidential campaign – defense-dependent companies, workers and communities have the clout to prevent reductions in military spending if they are not offered an alternative.

More broadly, the issue of R&D spending gets to the heart of the arguments for national industrial policies. It is increasingly widely accepted that military R&D spending has constituted a *de facto* U.S. industrial policy for the nation. Cuts in the military must be made with a high degree of national government consciousness of their R&D, industrial and global competitiveness effects. It is an open question as to how best the U.S. government should proceed with its nascent civilian-industrial policy. Should it reinstitute regulations to prevent transnational flow of certain kinds of technology? Should it offer tax incentives? Should it make direct grants to companies? Should it make substantial grants to states for them to retail to companies and communities?

Dual Use. The Clinton administration has addressed these questions through its Defense Conversion and Reinvestment Program, which bundles together a number of existing and new programs concerned with retraining opportunities, technology investment, and technology transfer activities. Probably the most relevant Clinton initiative for conversion of military R&D is the Technology Reinvestment Program (TRP), which is administered by the Advanced Research Projects Agency (ARPA) in the Department of Defense. Under the Defense Dual-Use Critical Technology Partnerships program, consortia of companies can apply for government funding on a cost-sharing basis to develop dual-use technologies; this popular program has attracted thousands of applications in each round of funding. TRP has also funded the creation of more than 35 manufacuring technology centers. It is worth noting, however, that total TRP funding is only a small fraction of the military R&D budget, or, indeed, of ARPA's budget.[29]

From an economic perspective, the TRP is intended to start the process of auctioning military R&D resources. Scientists often claim that military R&D teams are much more valuable if kept together as teams, although as we have seen theorists argue that the management of teams must change if new technology is to be transferred. In the TRP process, collaboration is encouraged and rewarded, because each partner is looking for another partner that will strengthen the collective application for funding. This ideally will combine the best of both worlds, bringing in new ways of civilian management while allowing military R&D teams to stay together. Thus in the marketplace of team-forming that occurs before each set of TRP applications, military R&D companies fill up their dance cards. This is a good exercise for the process of building new networks for civilian markets.

On the other hand, the TRP is supposed to have encouraged military companies to concentrate on certain civilian priorities, and this has not been happening quite as expected. For example, the TRP publicity made much of the fact that the TRP process would find high-technology answers to environmental and transportation problems. But fewer than 4 per cent of TRP projects are in these areas and they account for less than 1 per cent of the TRP funding.[30]

The Clinton administration has based its industrial policy on promotion of 'dual-use' technologies, referring either to basic 'building block' technologies like semiconductors, which are found in many industries, or to more specialized technologies that have definite defense and possible commercial relevance. The location of these programs within a defense agency, ARPA, however, raises questions as to how far the goals of conversion and building the civilian technology base will be served. Experts who claim that dual-use R&D is inherently futile have not been surprised by the precedence given to serving military goals. 'In the land of dual-use, the military side will always get first preference in terms of technology, scientists, and money.'[31]

Critics maintain that simply civilianizing ARPA would not make for a good institution:

> Nurturing technologies that are intended to succeed in the marketplace is more complex and exacting than supporting technologies for which government is the consumer. No matter how competent and motivated they are, there is every reason to fear that government officials, left to themselves, will be too far removed from technical substance and market feedback to make sound decisions.[32]

Dual-use incentives so far don't appear up to overcoming segregation between military and commercial R&D. This segregation is the reason the defense and commercial technology bases reinforce each other less than might be expected.

CRADAS. Beginning in the 1980s, the legal and administrative requirements for signing Cooperative Research and Development Agreements (CRADAs) between industry and government laboratories were progressively liberalized to promote technology transfer from the government to private industry. Like the TRP, CRADAs have attracted much interest from industry, with more applicants than available funds could support. By January 1995 the Department of Energy's three weapons laboratories alone had signed more than 300 CRADAs.[33]

These CRADAs provide for shared funding and access to research results. They provide a mechanism for the laboratories to contribute to the goal of enhancing national competitiveness, and for their scientists to make contact with their counterparts in commercial industry. These goals have been challenged, however, by the Galvin Commission, formed to advise the Department of Energy on the national laboratories, and the future of the CRADAs is in doubt.[34]

Other National Support of Corporate R&D. Reduced military R&D spending is a matter of concern because so much U.S. R&D has been driven by the Pentagon. For example, as part of its restructuring and downsizing General Dynamics reduced its R&D spending budget for 1990–1993 by $380 million, half the amount it spent the previous four years.[35] To avoid deterioration of national scientific resources, the federal government is being urged to take up the slack created by declining military spending by increasing its support of civilian R&D.

One way is to increase national incentives to encourage business investment in R&D. One of the biggest differences between defense contracting and civilian work is that civilian work has *no progress payments*. The civilian R&D must be financed through investment, which is why companies interested in R&D conversion have testified before Congress in favor of investment tax credits for R&D. Reforming the Pentagon's own practices can help. For example, the Pentagon's June 1992 decision to stop pursuing 'recoupment' fees for commercialization of Pentagon-funded R&D should benefit conversion within the aerospace industries. But it also has a worrisome side effect. The fees are imposed on U.S. companies' sales of both weapons abroad and commercial products that derive from military R&D. Thus, removing the fees may encourage conversion to commercial production, but it may also have the undesirable effect of increasing U.S. arms exports.

Another approach that would address the economic and political issues described above would be for the federal government to launch a major infrastructure campaign, with R&D components, offering incentives to states and companies to use freed-up military resources to invest in much needed social capital. This could be done by giving defense companies a kind of 'veterans preference' in bidding for contracts in the infrastructure campaign.

Actions by Private Companies

Among defense contractors, some companies are known for their R&D capabilities; if conversion is to take place it will have to occur in these companies . The companies prominent in the Strategic Defense Initiative are good examples. California's Hughes Aircraft Company appointed Bob Dankanyin as Senior VP for Commercial Diversification, probably the first in the defense business. His staff has coordinated 50 small diversification projects. Not all of these seeds are likely to bear fruit, but in 1992 Dankanyin checked off five he said were already producing profits for Hughes: (1) provision of satellite services; (2) wide-area networks that tie together far-flung business operations; (3) simulators for training; (4) air-traffic-control equipment and services; and (5) 'heads-up' instrument panel displays for cars. Other areas that looked promising to senior management include environmental services; projection systems for high-contrast, high-resolution video; commercial avionics; and radio-frequency ID systems.[36] Hughes tries to identify and then match core competencies of aerospace with market needs. Hughes' successes, however, require the deep pocket of a large corporation. For smaller companies, state and federal assistance is required in the form of (1) technical assistance for strategic planning and (2) financing for feasibility studies, marketing and actual projects.

In Maine, Bath Iron Works - long kept going by defense contracts for naval warships - is attempting, with the help of Maritech (a defense conversion program operated by the Department of Defense), to reconstruct itself along the lines of Japanese and Finnish shipbuilding companies. The makeover is being undertaken with union support and a portion of nearly $200 million of FY 1994 federal aid for civilian shipping companies.

Universities

Since the 1950s, the U.S. Department of Defense has contracted with universities to conduct military R&D, to the distress of some within the groves of academe. The universities' dependence on the Pentagon was an ethical Faustian bargain, trading encroachment on academic independence in return for the military engine that allowed them to build new scientific and engineering facilities and hire new faculty and staff. It may also be an economic Faustian bargain, as the engine has stalled and - if the Republican Congress stresses balanced budgets over military strength - may go into reverse. Universities that have depended on DOD funds to support facilities and graduate students will find it difficult to find alternative sponsors at the same level of funding. The other side of this coin is that graduate students in science and engineering were encouraged to specialize in areas of interest to the Pentagon as opposed to private industry. Federal assistance is needed to encourage students to specialize in the civilian technologies of the 21st century, thereby building our international competitiveness.

State and Local Programs

In the United States, state and local governments play an important part in industrial policy. Regional mechanisms for converting military R&D are very promising. A broad-scale regional program typically will include such features as a small-business technology incubator, informational clearinghouses, marketing fairs, training programs. Some examples of programs directed at conversion are:

- California's Regional Advanced Manufacturing Project (RAMP) was established as a three-year partnership between companies, unions and the California State University System and Community College System to seek new civilian markets and the most efficient methods of production for them. The project has, however been weakened by delays in funding that led to the withdrawal of two key partners.
- Connecticut was one of the first states to recognize the importance of community-based, as opposed to company-based, planning for conversion, and the state has a number of interesting programs including loans for conversion activities. In May 1994 Connecticut passed An Act Concerning Economic Development Program Accountability, which makes job retention a priority in economic development and provides for planning, oversight and penalties to ensure this is the case.
- Nevada's Senator Richard Bryan, working with national conversion advocates, obtained $10 million to start research on solar energy production and solar-cell manufacturing at the Nuclear Test Site northwest of Las Vegas. This project appears to be progressing well.
- New York State was late getting started, given the magnitude of its defense-dependence on Long Island, but with some federal assistance it has launched a useful study of the potential for civilianization of defense contractors and subcontractors on Long Island. The study has made a significant difference in restructuring 10 firms to enhance their commercial prospects.

INTERNATIONAL ASPECTS

In this section we review the obstacles to and mechanisms for conversion of military R&D in the international environment.

Competition as an Obstacle

Military R&D has been a driving force behind the arms race. The end of the Cold War and subsequent drawdown in military budgets has led many governments to view arms exports as a way of maintaining their defense industries.

The Russians in particular have viewed foreign sale of their weapons as a way of meeting their urgent needs for hard currency. They have run into competition from other arms suppliers who are selling higher-technology weapons and have a competitive advantage. Without effective controls on the arms trade, the arms sellers will continue to argue for continued investment in military R&D to provide a competitive edge.

Cooperation as a Mechanism

More optimistically, we can find reasons to expect cooperation among countries. Cooperation should be possible because the problems of conversion are universal. The search for alternative activities for defense enterprises goes on worldwide. The disdain that highly trained scientists have for working on consumer products knows no borders. The corresponding disdain by consumers for the overly complex design of consumer products by defense engineers is also an international phenomenon.

Using 'Nunn-Lugar' funds, first appropriated in 1991, the United States is helping to finance many projects in the republics of the former Soviet Union to help these new countries civilianize their military research and production. A number of joint ventures have been undertaken between the United States and military R&D institutes in Belarus, Russia and Ukraine. Among the joint projects that have been initiated are:

- Manufacture of surgical instruments (U.S. company Baxter International with the Russian Research Institute for Automation and Instrument-Making, 1993).
- Low-temperature manufacturing of diamond-coated drill bits (U.S. National Patent Development and Kurchatov Atomic Energy Institute, 1992).

- Manufacture of shaving equipment (Gillette and Leninets, 1993).
- Conversion of SS-18 and SS-19 missiles to launchers for communication satellites (Motorola and Krunichev Enterprise, 1992).
- Improvement of nuclear power plant control and safety (Westinghouse with Kharton Production Association in Kharkiv, Ukraine, and with RAO Rossi in Russia, 1994).
- Improved avionics in Russian planes (Allied Signal with the National Institute of Airborne Avionics Equipment, 1992).
- Improve air traffic control systems (Rockwell International with the State Research Institute of Aviation Systems, GosNIIAS, 1994).[37]

The advantages of cooperation with the newly independent states include helping to lock in conversion there, making the best possible civilian use of their technological strengths and top scientific personnel. It also has an important symbolic value, enabling implementation of disarmament treaties with confidence that both sides are committed to movement toward a more peaceful world. One approach to this would be through international lending institutions.[38]

Possible Roles for the United Nations

The United Nations has sponsored several conferences on economic conversion of military production. Is there more that the U.N. could do for conversion, especially of R&D? Here are some suggestions:

- Assess the world's civilian R&D needs. A panel of experts could establish some consensus on global priorities for R&D such as energy sources and conservation, environmental preservation and disease control.
- Seek to establish *international* R&D labs to pursue civilian research on world health and ecological threats. Scientific projects might be organized internationally, which might be a way of keeping together teams that would otherwise be dispersed. Also, the political decision on the need for irreversibility might be different if asked on the international level as opposed to the national one.
- Seek to limit global military R&D spending.
- Expand its monitoring/verification activities toward global reduction of weapons of mass destruction.
- Support the proposal of former Costa Rican President Oscar Arias Sanchez, the 1987 Nobel Peace Prize Laureate, for an international fund for peace.[39]

Forging an International Consensus for R&D Work

Since the nationalistic motivations for military R&D appear to have been undermined by the end of the Cold War, a new basis for public support of R&D is required. Recapturing the kind of consensus that emerged for wartime and Cold War research on weapons may depend on international reorientation of both tasks and research teams - either on a bilateral, regional or global (U.N.) basis. Consensus is most likely to grow around R&D projects that have long-term benefits for the greatest possible number of people on the planet.

CONCLUSIONS

The main obstacles, in the United States and elsewhere, to converting resources now consumed by military R&D to civilian uses are a lack of information about such spending; regional economic dependencies; the isolation of military R&D from civilian markets; special interest lobbying; and the difficulty of retraining military R&D specialists; and, especially in nonmarket economies, the lack of alternative markets.

The most promising mechanisms for converting military R&D are: (1) improved information on military spending, especially on R&D, worldwide, (2) civilian university programs to absorb the loss of military ones, (3) in-house diversification projects by military contractors, (4) regional conversion programs, and (5) national initiatives, such as a proposed campaign to rededicate some military R&D resources to improving infrastructure elements.

Finally, we conclude that competition among arms producers for foreign sales remains a major obstacle to conversion and a dangerous source of global instability. This obstacle must be addressed by international bodies, using cooperative mechanisms to spell out global civilian R&D priorities and provide for collaborative programs to achieve these goals. The proposal for an international fund for peace should be seriously considered.

NOTES

1. For collaboration and helpful suggestions over many years the writer thanks especially Domenick M. Bertelli and Jurgen Brauer, Hugh DeWitt, the late Admiral Stanley Fine, Alice Tepper Marlin, and Judith Reppy.

2. Maggie Bierwirth, 'Capitol Hill and Conversion: A Summary of Recent Congressional Action,' in *Real Security: Converting the Defense Economy and Building Peace*, ed. Kevin J. Cassidy and Gregory A. Bischak (Albany: State University of New York Press, 1993), and 'Defense Reinvestment and Conversion,' *The New Economy*, Winter 1994.
3. Defense conversion in the Department of Commerce appropriation (Economic Development Administration) at this writing is budgeted at $36 million, only 30 per cent of the $120 million requested by the Clinton administration and a $62 million drop from the amount approved by the House Appropriations Committee. The Manufacturing Extension Partnership received $51 million, a $30 million drop from the House and a $40 million drop from the administration request. President Clinton reportedly indicated that he will veto the Defense Appropriations bill in an attempt to shift the $7 billion in defense spending above his request to job training and education. Source: Jim Bridgman, 'Conversion Legislative Update,' September 27, 1995 (Washington, DC: National Commission for Economic Conversion & Disarmament, Internet distribution).
4. National Science Board, *Science & Engineering Indicators–1993* (NSB 93–1), Washington, D.C.: US Government Printing Office, 1993, p. 315.
5. Albert H. Teich, 'An Introduction to R&D in the Federal Budget,' in *AAAS Report: Research & Development, FY 1996* (Washington, DC: American Association for the Advancement of Science, 1994).
6. Seymour Melman, *The Permanent War Economy: Capitalism in Decline* (New York: Simon & Schuster, 1985); Jacques S. Gansler, *Affording Defense* (Cambridge, MA: MIT Press, 1989).
7. In the course of conducting defense-industry interviews in 1992, the Council on Economic Priorities found that companies like Boeing were observing this separation less rigorously than in the past because of the need to move personnel.
8. Joshua Lerner, 'The Mobility of Corporate Scientists and Engineers between Civil and Defense Activities,' *Defence Economics* 3:3 (1992), pp. 229–42.
9. John A. Alic, Lewis M. Branscomb, Harvey Brooks, Ashton B. Carter, and Gerald L. Epstein, *Beyond Spinoff: Military and Commercial Technologies in a Changing World* (Boston: Harvard University Press, 1992), p. 375.
10. Alic et al., *Beyond Spinoff*, p. 108.
11. Mark Green, ed., *Changing America: Blueprints for the New Administration* (New York: Newmarket Press, 1992), p. 762. See also Kenneth Mayer, *The Political Economy of Defense Contracting* (New Haven, CT: Yale University Press, 1991).
12. Peter Passell, 'High-Tech Industry is Hard to Help,' *New York Times*, 2 February 1993.
13. Alic et al., *Beyond Spinoff*, p. 9. The economic value of the spinoff from the high-tech Strategic Defense Initiative has been challenged in Rosy Nimroody, *Star Wars: The Economic Fallout* (Cambridge, MA: Ballinger, 1988).
14. John Markoff, 'Building the Electronic Superhighway,' *New York Times*, 24 January 1993.

15. *Adjusting to the Drawdown*, Report of the Defense Conversion Commission, December 31, 1992, p. 24.
16. Lewis M. Branscomb, 'Does America Need a Technology Policy?' *Harvard Business Review*, March-April 1992, p. 24.
17. Warren D. Siemens, 'Partnerships with Industry and the Oak Ridge National Laboratory,' in *Technology Transfer in Consortia and Strategic Alliances*, ed. David Gibson and Raymond Smilor (Lanham, MD: Rowman & Littlefield, 1992), pp. 158–59.
18. Michael S. Rubin, 'Creating and Managing Consortia Cultures through Transitional Episodes,' in Gibson and Smilor, *Technology Transfer*, pp. 70–71.
19. Bill Curtis, 'Technology Transfer in Knowledge-Intensive Organizations,' in Gibson and Smilor, *Technology Transfer*, p. 33.
20. Claude DelFosse, 'Technology Transfer at the Software Productivity Consortium: A Partnership Approach,' in Gibson and Smilor, *Technology Transfer*, p. 113.
21. Debra M. Amidon Rogers, 'Science as a Competitive Weapon: Utilizing Alliances in Global Markets,' in Gibson and Smilor, *Technology Transfer*, p. 43.
22. Wendell M. Fields, 'Technology Transfer in a Diverse Corporate Environment,' in Gibson and Smilor, *Technology Transfer*, p. 63.
23. Marie R. Jones, *Converting the Cold War Economy: 63 Companies Change for the Future*, report of the Center for Economic Conversion, August 1992.
24. Regulatory policy has perhaps been more successful, as AT&T, for example, has continued on the breakup path started for it by the Justice Department.
25. Branscomb, 'Does America Need a Technology Policy?' p. 27.
26. Steve Lohr, 'Lessons for a High-Tech President,' *New York Times*, 24 January 1992.
27. The Internet address (Web site) is: http://ecix.doc.gov.80/ecix/ecixhomepage.html.
28. Internet address: http//bicc.uni-bonn.de.
29. See Defense Budget Project, 'Technology Reinvestment Project: Potential Military Bargain,' 17 February 1995. For a less positive view of the TRP, see Michael Oden, Greg Bischak, and Christine Evans-Klock, *The Technology Reinvestment Project: The Limits of Dual-Use Technology Policy* (Washington, DC: National Commission for Economic Conversion and Disarmament, 1995).
30. Interview with Erik Pages, National Economic Conversion Information Center, U.S. Department of Commerce, November 9, 1994.
31. Seymour Melman, quoted in David Evans, 'Dual-Use Technology Could Prove to be a Troubled Marriage,' *Chicago Tribune*, 15 January 1993.
32. Alic et al., *Beyond Spinoff*, p. 375.
33. Andrew Lawler, 'Report to Stress Research Over Close Ties to Industry,' *Science* 267 (27 January 1995), pp. 446–47.

34. Department of Energy, Secretary of Energy Advisory Board, Task Force on Alternative Futures for the Department of Energy National Laboratories, *Alternative Futures for the Department of Energy National Laboratories*, February 1995. The Chairman of the Task Force was Robert Galvin, Chairman of the Executive Committee of Motorola, Inc. The full Galvin Report is retrievable through the Internet on the Department of Energy home page. For another viewpoint, see William J. Weida and Ann Markusen, 'An Alternative to the Galvin Report on Futures for the DoE Nuclear Weapons Laboratories,' Rutgers University Center for Urban Policy Research Working Paper #87, 1 February 1995.
35. U.S. Congress, Office of Technology Assessment, *After the Cold War: Living With Lower Defense Spending*, OTA-ITE-524 (Washington, DC: U.S. Government Printing Office, February 1992), p. 199.
36. Interview with Dankanyin, February 1992.
37. List prepared by Anna Gincherman, Council on Economic Priorities, August 1994.
38. Jordana Friedman and Mulan Ashwin, 'Aid as a Tool for Arms Reduction,' *Research Report*, Council on Economic Priorities, March 1995.
39. This proposal has been intelligently amplified by Michael Renner in *Budgeting for Disarmament: The Costs of War and Peace*, Worldwatch Paper 122 (November 1994), pp. 50–56.

7

The Conversion of Military R&D in the Russian Arms Industry

Mikhail Nikitin

INTRODUCTION

The solution to the problems of converting military science depends upon the ability of the national economy to make effective use of the technical and intellectual potential concentrated in military industry. It depends also upon any measures taken by the state to preserve the existing level of social protection for scientists, technicians and workers and to stimulate speedy structural change in this branch of industry. Finally, it depends on the laws regulating the orderly integration of the converted science and production into the civilian economy. In the transitional stage of conversion from a centrally controlled economy, characterized by a total monopoly on property and most kinds of resources and by the absence of economic mechanisms such as taxes, profits, prices, credit and the like, we cannot assume that the process of converting military science will be rapid and effective.

In this connection we must consider the degree of conversion of military science already carried out in Russia under the old rules, since the results will have a continuing influence on the general direction of the conversion process. The first stage in perestroika, the program for general restructuring of the Soviet economy, was the attempt in 1987 at massive conversion on the basis of a centralized state program. The plan was to utilize the potential of military science for raising the technical level of branches of heavy industry in the creation and production of technical equipment for the agricultural industry and new automated lines and equipment for the textile, clothing and food industries.

The program had a couple of drawbacks.[1] First, it was never published, not because of any special secrecy, but because the bureaucrats in charge of it were afraid that they would be severely criticized by experts if they made their plans known to the public. Second, it was mostly a top-down approach, seldom based on the conversion plans of individual plants, although Gorbachev had promised in his December 1988 address to the United Nations in New York to deliver to the world public a number of such plans before the end of 1989. This reflected the usual Soviet preference for macroeconomic programming, which was more compatible with managing the economy within the existing system. Third, the program allocated money - a considerable amount - for the conversion exercise, but it was not earmarked and so never reached the enterprises. Fourth, the whole economic system was already disrupted, so one of the most important problems for conversion, that of investment, remained unsolved.

Large funds were allocated to this program. By exploiting the scientific potential and experience of the defense ministry in the production and use of computer techniques, and through the central provision of material resources and techniques, a definite basis was created for raising the output of the civilian engineering industry and also its technical level. But, on the whole, this stage in conversion was focused on very specific projects. It did not represent the transfer of available scientific military potential to civilian purposes, but rather the injection of new power into the civilian sector of the economy on the principle of organizing isolated production facilities under the direct control of the defense ministry. This stage of conversion gave no positive results for effectively exploiting the scientific and technical potential of the military science complex. One of the results of this type of conversion proved to be the additional cost of using the military science base to solve the problems of raising the technical level of civilian industrial production, with unavoidable consequences for the maintenance of the Soviet economy at the end of the 1980s.

The second stage in the conversion of military science was a directive issued in 1989 that reduced the budget allowance for purchasing arms, military technology, scientific research (NIR) and experimental design work (OKR). This stage took place at the same time as the concept of the market economy was being introduced and along with new legislation covering the manner in which the new economic policy of the state would operate. The ensuing dramatic fall in military production and the rupture of established economic links brought military science to a critical situation. Military science establishments, being still the property of the state, were under state jurisdiction, but they were practically deprived of support. Qualified specialists in military science began to transfer to small businesses or become entrepreneurs. This stage also signaled the beginning of a loss of prestige for scientific and engineering work.

During this whole period, right up to the middle of 1992, military science in Russia, with its high intellectual potential, was in a state of complete uncertainty as to the demands from the state and from society. This stimulated the military science establishments to undertake a wide-ranging search for new ways of utilizing their accumulated scientific and technical potential and to study what demands the market might make on their scientific expertise. Meanwhile, the center of the reconstruction process moved into the sphere of politics, with the formation of new state management structures, the development of ideas about the mechanism of relationships within the market economy, and the introduction of normative and legislative measures underpinning these relationships. The most important aspect of this structural policy involves eight priority activities, one of which is conversion. In essence, one is talking not so much about conversion as about a process of reorienting defense technology towards the needs of the civilian economy, that is, the reform of the whole military-industrial complex to meet the demands of the new defense doctrine of the Russian state. Restructuring the defense industry, however, ought to be implemented not only at the macroeconomic level, affecting branches of the defense industry as an interrelated complex of science and production, but also at the microlevel, where the economic strategy and tactics of development affect the establishment itself.

Only in 1992, with the passing of the law 'On the conversion of the defense industry in the Russian Federation' did Russia formulate a definite attitude to macro- and microeconomic concepts and define a series of legal norms for the state policy on conversion covering the principles of state protectionism, economic regulations and programming, shareholding, and privatization. A system of priorities was set up for the possible uses of the scientific expertise of establishments liberated for conversion. The basis of conversion became the programs and projects fixed by the state for the converted establishments, based in turn on their business plans. These programs and projects were now to be financed from various sources, depending on the interest shown by investors in the commercial or other outcome of particular projects. The source might be the state, a commercial undertaking, the bank for innovations, a foreign investor, the establishment itself, credits received, and so on. Another important source of funds for conversion projects might be municipal authorities interested in solving regional aspects of development.

The economic policy introduced by the Russian government and underpinned by a system of laws and regulations already corresponds, to a large extent, with generally accepted principles of market economics, and it has stimulated the conversion process and allowed establishments to fit into the civilian labor system. Thus, by the end of 1992 the conversion processes had been given a system of points of reference set by the state; the economic mechanisms for stimulation and regulation had begun to function; and provision had been made for legal support in the selection of the form and direction of organized scientific and production activities. Questions of shareholding and privatization of defense establishments, including military science, and the organizational, legal and economic aspects of conversion of military science and production were by this time fairly well defined.

Conversion possibilities depend in part on the structures of the enterprises performing military R&D. In the arms production system there are two types of scientific and manufacturing subdivisions in the industrial establishment - independent NII (scientific research institutes) and KB (design bureaus), and research and manufacturing subdivisions (self-financing subdivisions). The activity of these and other actors was specialized along strictly defined lines of military or civilian techniques; duplication of work when devising military techniques was, as a rule, prohibited. Narrowly-specialized research areas were explored by staffs of qualified specialists in the various research organizations. There is, of course, a definite advantage to a system in which the scientific products created are implemented within the branches of the same establishment, and even within the defense establishment as a whole. Under these conditions we have a scientific-research and manufacturing subdivision included within the structure of the industrial establishment. As a result of this tightly linked structure, however, the conversion of military science is possible only if questions of the conversion of military production are considered at the same time. These were the initial conditions for the lengthy processes of conversion.

CURRENT CONVERSION EFFORTS

How are the actual processes of conversion working? Let us take as an example the establishment, 'Main Administration of the Arms Industry: The Committee of the Russian Federation for Defense Branches of Industry' (GUPVKRF). Before the beginning of conversion the cost of military research and design work was from 6 to 24 per cent of the price of various types of armaments and technology development carried out by establishments of this branch of industry. By 1992 more than half of all establishments had converted to civilian production of at levels of 50 to 70 per cent, with a minimum level of conversion of 19 per cent and a range up to over 87 per cent of the workers in establishments. The state, however, compensated only 10 to 20 per cent of the scientific-productive potential that was set free, through orders for the creation and production of techniques for civilian use; moreover, orders for the completion of military research and design work did not exceed 20 per cent of the potential output of these establishments. Thus, there was no demand for more than 50 per cent of the scientific productive potential of arms-manufacturing establishments belonging to the state. This kind of conversion gave rise to a sharp reduction in the profitability of the military production and to structural distortion. The scientific research and production subdivisions in effect lost the fundamental basis of material production that enabled them to sell their scientific-technical product. Only the establishments specializing in opto-electronics, which had a higher degree of assimilation and much higher science content in the final product, could really begin to adapt to the conditions for conversion laid down in 1992.

What is typical of the conversion process at the present time? Above all, these conversion processes have the common tendency to reduce the degree of militarization of the economy in Russia to a level set by the doctrine of defense sufficiency. The consistent efforts of the government of the Russian Federation in reducing the level of international confrontation and in the agreements on arms reduction makes the conversion process irreversible. This has destroyed the 70-year-long principle built into branches of the defense industry - the constant increase in military technological potential to serve the security needs of the state - and, as a consequence, the need for complete control by the state of the finances of organizations and subdivisions of military science.

The high priority allocated to defense conversion and the corresponding legal measures of support for this policy have accelerated structural changes in the defense industry and stimulated establishments to diversify and to make use of their accumulated scientific and technical potential for civilian applications. Diversification of military science is characterized by participation in the development of joint projects by most of the state enterprises, these projects being designed to fulfill the state program for the creation and production of civilian techniques, goods and services, as illustrated in Table 7.1. In order to stimulate the conversion process, the state underwrote favorable credit terms for the implementation of conversion programs by defense establishments in NII and KB. However, the amount of these credits in 1992 covered less than half the cost of the structural change demanded by those establishments.

In 1992 the NII and KB, together with scientific, technical and manufacturing subdivisions of industrial establishments, proposed a wide range of projects for applying high technology to civilian applications. In branches of the arms industry alone these proposals took the form of more than 400 different imaginative projects, such as the creation of equipment for neutralizing industrial effluent, deep pumps with low energy requirements, refrigeration systems, production lines for baby food and pharmacological preparations, therapeutic equipment, diagnostic equipment for the fetus, means of transporting potable liquids, multiple-use tractors, mobile generators of up to 350 KW output, light bulletproof waistcoats, lathes for turning wood and stone, drills with internal cavities for cooling, dosimeters for ultraviolet radiation, infrared cameras, binoculars with high resolution, apparatus for night sight, spectacle optics, laser-operated compact disc players, photographic equipment for monitoring air transport, microwave ovens, turbocompressors, kilns for baking ceramic tiles, machines for monitoring knitted cloth, and so on. Up to the present, however, there has been considerable difficulty in implementing these conversion projects. The market demand has been insufficient to utilize the accumulated scientific and technical potential. The projects chosen for action did not employ the full capacity of the personnel of the scientific, productive and experimental subdivisions of NII and KB. In a series of projects not guaranteed by the NII and KB, up to 40 per cent of the staff did not participate, particularly in the experimental subdivisions, which are often fitted with costly and unique equipment.

There are serious financial problems. The high level of value-added tax depresses the demand for high-tech products. The inflation index coefficient allowed by the state finance department for firm orders of goods required by the state does not correspond with the actual growth of prices, and the insufficient amount of credit for complete work complicates the financial position of factories undergoing conversion and fails to protect the social security of workers in military science, even at a reduced level. Support for the infrastructure of converted establishments is practically non-existent. Finance for associated military research and design work is subject to considerable interruptions and to long delays in the payment of bills by the customer at each completed stage. The average level of workers' salaries in the converted military science establishments that remain government property is from one half to one fifth the salaries of specialists in the state department for control systems, which has budget support, not to mention the salaries in commercial organizations.

Table 7.1
Participation of establishments belonging to the arms industry in priority programs of conversion, 1992–1995

Name of the program	Number of establishments participating	Most important conversion projects
Conversion-transport and communications	36 industrial factories and unions	Creation of substitutes for imported goods in view of the disintegration of the USSR, e.g., production of oil tanks, heavy goods vehicles, and dump trucks with high mobility.
Conversion-agricultural & consumer sector	44 industrial factories, 27 NII & KB	Production of mini-tractors and powered hoists. Production of consumer goods, e.g., vacuum cleaners, refrigerators, food processors, video recorders, etc.
Conversion-medical	17 industrial factories, 20 NII & KB	Production of cardiac stimulators, complete sets of medical instruments, diagnostics apparatus.
Conversion-ecology	4 industrial factories, 13 NII & KB	Spectroanalyzers for checking purity of the environmental media and of raw materials, electromembrane desalinators, etc.
Conversion-chemical-timber complex	10 industrial factories, 3 NII & KB	Timber-hauling machines, complex treatment of wood waste.
Conversion-fueled energy complexes	13 industrial factories, 18 NII & KB	Gas turbine compressor equipment for compressing natural gas. Complex oil equipment for fracturing strata.

All the above leads to instability in the activities of converted establishments at this important stage in carrying out the conversion process. There is a loss of prestige for work in this sphere, which enhances the 'brain drain' of qualified staff. More than 15 per cent of the scientific and engineering personnel of converted establishments have left the service during the last four years. This tendency is cumulative, and if insufficient attention is paid to the problem it will result in a serious loss of scientific potential for military science. The threat of wastage of scientific potential for military science is due not only to a shortage of orders for military research and design work, but also to some aspects of the monetary economic policy of the state, and to the absence of a well-thought-out investment policy. Investment famine also affects the reduction in orders for scientific programs in the civilian sector of the economy.

To some extent the brain drain of specialists as a result of conversion on a national scale is natural and proper, accompanying the formation of a free market for labor. However, under the conditions obtaining for conversion in Russia, this brain drain may have a negative effect on the intellectual potential of the country as a whole. In the absence of demand for their scientific and technical work specialists may end up, more often than not, in purely commercial jobs where, in the course of time, they become exhausted for lack of due recognition by the state. At the same time this outflow of scientific specialists stimulates and supports a new tendency - viz, a reduction in the number of young people striving to complete a course of higher technical education - and this will result in a lowering of the general intellectual potential of the country.

Despite all these problems, diversification of the basic activities of establishments that are oriented towards military work has proved to be an effective method of conversion. Experience with the diversification of Russian establishments during the initial stages of conversion (1990–1992) and, in particular, with NII and KB, has shown that in the scientific research and manufacturing subdivisions of industrial establishments concerned with arms manufacture the success of diversification was very closely linked with the nature of their basic activities. We shall illustrate this by the case of the scientific-production complex formerly engaged in building the strategic missile SS-20.

The Votkinsky factory, having ceased work on the SS-22, and having earlier experience in making children's perambulators, now began also to produce large refrigeration equipment with complex control systems, washing machines, microwave ovens, and very precise metal-cutting equipment. The scientific research institute for the design of strategic missiles, which before the start of conversion had practically no civilian scientific activities, has turned much of its scientific strength towards basic applied research and development, thus encouraging progress in:

- equipment and machines for the agricultural industry
- improvements in the machine construction field
- medical technology
- consumer goods such as food mixers

- new materials
- special technological equipment for sections of the machine construction industry
- systems and means of information
- production and testing of models for some of the scientific projects outlined above
- modular factories for working flax
- a multifunctional mobile complex for emergency help and rescue at sea
- exploratory work on incubators for pregrowth of baby chicks
- preliminary work on the creation of industrial facilities for the cultivation of vegetables.

The wide range of scientific themes listed above required the retraining of a considerable number of the institute's specialists, as well as reequipment of the experimental and testing divisions. These researches and projects were carried out within the planned conversion program, financed in part from the budget and favorable credits distributed under the priority programs. Some projects were undertaken by the institute by means of credits from commercial banks. The Principal Arms Industry Committee of the Russian Federation, which manages the complex that includes the scientific institute and the Votkinsky machine construction factory, was guarantor of credits for the foregoing projects.

Another example of structural change on the micro level of the defense-oriented scientific complex is provided by the Central Scientific Research Institute for Economics, Information Theory and Systems Control, formerly entirely devoted to solving problems in the field of automated systems of production control in defense industry establishments (ASUP). The need to work out typical design decisions in the field of ASUP and systems of automated design of the CAD/CAM type and to complete work on economic mechanisms for the management of defense establishments and of NII and KB left very little freedom for creative investigations by economists, systems engineers, programmers and technologists under the conditions of state control outlined above. Such themes fitted well, however, with the structure of scientific subdivisions closely oriented towards tasks set by the principal client – the Ministry of Defense Production.

The change from centralized systems of control of the branches to a market system of management and primarily economic methods of state regulation also changed the direction of the scientific departments in important respects. Henceforth their research and development work became primarily oriented towards satisfying the demands of the market, and they were intimately involved in solving the problems arising. In this connection also entirely new subdivisions were formed, for example, a center for marketing research. Or - another example - a center for strategic research to work out the direction that scientific policy should take under the new defense doctrines and foreign policy initiatives of the Russian government. Within the institute a center of expertise for innovative programs and projects was established with the duty of evaluating projects in order to decide whether to invest in them. An important part was played under the new conditions by the center for privatization and antimonopoly policy, which organized and underwrote the privatization and sale of shares in defense establishments. In general, the institute represented a cooperative of economically independent scientific centers, and the institute itself was reorganized into a holding company.

Thus, under the influence of the conversion program a structural policy took shape in the branches of the arms industry. However, the efficiency with which conversion and the associated diversification is carried through depends directly upon fundamental structural change in the whole war-oriented industry. Experience with economic reform in Russia shows that the attempt to convert the Russian military industry without restructuring the whole economy is doomed to failure.

INTERNATIONAL DIMENSIONS OF CONVERSION

The scientific potential of converted enterprises is not limited to national programs and projects. Among the international aspects of conversion one must include lowering the level of confrontation between opposing blocs. This level is set by political decisions. At the present there is a real foundation for a maximum reduction in armaments, especially the nuclear component. The first steps have been taken in this direction - agreement has been reached on an important reduction in the nuclear capacity of the two leading nuclear powers - and one may say that peaceful cooperation has begun in this field. Reducing the nuclear confrontation demands great efforts and mutual intellectual participation if this process is to proceed without lowering the national security of all countries.

Another of the international aspects of conversion is to get countries that are converting their war potential to participate in a system for the management of peace - to play their part in a peaceful system of division of labor. Such participation must be regarded as utilizing the liberated intellectual potential for the solution of world problems affecting the whole of society: We already have examples of such cooperation, and they must be extended.

We need to examine joint participation in defense projects and programs, after the manner of the earlier separate projects of the NATO countries and of the Warsaw Pact. However, the foundation of such projects should not be the production of weapons, but the use of high technology ('know how') and the deployment of joint scientific staffs in combined scientific centers, laboratories and institutes. One important direction in which such joint activity could take place is in the application of defense techniques to civilian problems.

Clearly, the highest form of collaboration in applying converted scientific and technical potential might be in setting up a joint study of the directions in which science and technology should be developed. It is well known that the general practice of conversion is founded on the principle 'do more research and less production.' Holding to this principle creates more preconditions for cooperation in defense technology, and what is important, in the convergence of high technology coming from scientific research, whether developed for defense or for civilian purposes.

It seems that the path of national conversion ought to follow worldwide experience in the same direction, for then new ways open up for closer international integration: for example, in creating international norms for broadcasting and the use of binary technology in the interests of the worldwide community; coordinating of defense doctrines of the various states while resolving the problems of national security; working out joint programs for training staff for the solution of important conversion projects; developing programs for the joint use of high technology ('know how') in the interests of the world community; and in the creation, development and use of data banks in the field of binary technology; and so on. At the present time Russia is ready, as never before, to put forward a wide range of proposals in the interests of international collaboration in science and technology. Thus, from establishments of the GUPVKRF alone, an information bank has been formed containing more than 260 projects for new techniques and equipment in manufacturing industry and 150 scientific studies, more or less ready to be applied, of advances in the fields of medicine, ecology, transport and the like. Some of these projects are already being implemented with the participation of partners from abroad. For example, there has been cooperation between the Leningrad (now St. Petersburg) opto-mechanical union and the Japanese firms Olympic Optical and Iskra for developing and producing flexible endoscopes for medical use. Other examples are the collaboration with the firm 'Dzhuky' for the production of industrial sewing machines, and with the firm 'Konica' for manufacturing cameras and mini-laboratories. Many more possibilities for joint conversion programs have opened up under the new Russian regulatory legislation in the field of shareholding and privatization of establishments in the defense complex that allows foreign investors to participate.

The way forward for conversion in Russia by international collaboration must not be confined to attracting foreign investment, and even less should it be limited to receiving foreign credits. Conversion of the defense industry opens up the possibility of implementing joint projects, both inside the country and abroad, by exporting high technology and scientific potential in both the military and the peaceful sectors of the economy. The initial stages of such collaboration could be organized on the basis of regional conversion programs involving investment by the countries concerned in achieving the relevant projects.

If a larger measure of international collaboration could be achieved one might undertake a joint project for building an adequate communications system in Russia. Over many years the information media in the Soviet Union were built up under the constraints and prohibitions imposed by the regulatory regime imposed by the western bloc through the Coordinating Committee on Multilateral Export Controls (COCOM). As a result of this policy, not only Russia but the whole world community has been deprived of normal communication links and access to many treasures of literature, art, science and technology. Intellectual property in many branches of knowledge in the Russian Federation and the former republics is at present the victim of speculative and illegal business, and Russia is practically without up-to-date information media. Even the current expansion of information media to deal with the move to a market economy does not resolve the problem. Rather, it is reminiscent of the practice of supplying Russia with old-fashioned techniques.

The intellectual potential of Russia in the field of information media has not yet been exploited to the limits of its competence and possibility, in spite of the huge demand for modern and reliable information products in every sphere of activity. International collaboration in this field of scientific industry could serve not only for the solution of problems in information technology, but also as a basis for improving the world political climate. Timely and reliable information is the best source of mutual understanding and trust between countries. This collaboration could become the foundation of a global international information network and also the basis for realizing the intellectual potential and productivity of converted industries.

It is well known that the spread of information through the world community is the most effective path for progressive development in all branches of science and technology, industry and social infrastructure, and ecology and protection of the environment. This path of international collaboration is so many-sided that it would undoubtedly prove to be a source of many individual projects of collaboration between converted industrial and scientific research firms and organizations in all spheres of science-based activity.

CONCLUSIONS

The actual political and economic conditions within our country do not as yet encourage much interest in the developed countries in realizing the proposals outlined above. There is obviously still a definite danger of instability in the reforms that have taken place in Russia. And the continuing effect of stereotypical confrontational thinking is perhaps not eliminated in other countries. It is possible that opinion in these countries considers it more expedient to regard conversion of Russian scientific and technical potential as a 'brain drain.' There is, however, no doubt that if the Russian government shows a preference for such forms of conversion it must urgently seek ways of preventing the 'drain.' It must also avoid economic development proceeding along paths that would assign to Russia the position of a supplier of raw materials to the developed countries, with a predominance on its own territory of ecologically dangerous products involving the primary treatment of resources using a high level of manual labor.

Russia is the first country to begin far-reaching conversion of military science and production, and the world community should appreciate its efforts to demilitarize the economy and should accord it a worthy place in the world system of the division of labor. It seems that the solution of the problems of conversion should be accomplished within the framework of the world community through an international program of intellectual conversion of military science in which all interested countries would participate. A special committee of the United Nations would have to draw up and coordinate such a program. The world community must exclude the possibility that the scientific conversion achieved could be utilized for the production of new types of weapons or military techniques. If an arms race were to occur within the frame of national state programs or international programs of alliances, then it would be necessary to construct a system of general security based, it seems to us, on a program common to all countries of the world - an international program of conversion.

NOTE

1. This paragraph is based on material supplied by Yuri Andreev.

8

Conversion of Military R&D: The French Case

Jean-Paul Hébert

INTRODUCTION

This chapter addresses the issue of the conversion of military research and development (R&D) in France. Various aspects are examined: 1) The importance of military R&D; 2) the extent of de facto conversion of French arms industries; and 3) the specific position of military R&D in the French armament policy and the difficulties in carrying out R&D conversion.

French military R&D is not clearly defined: the government research department's annual assessments regularly emphasize that the estimates are approximate. Moreover, the aggregates used to measure R&D (RDT, Research-Development Tests; or MRDBE, military R&D budgetary expenditures) do not match international criteria (particularly those of OECD). Taking these inadequacies into account, it can, however, be estimated that military R&D (MRDBE), which was 36 billion francs in 1991, represents more than one-third of public R&D expenditures. Since 1983 the share of R&D in the defense budget has continually increased from 13.6 per cent of the budget in 1983 up to 18.5 per cent in 1991, which is the highest percentage of all the OECD countries. In absolute terms, the total amount of French military R&D is second only to the United States among OECD countries. French military R&D is thus important for its budgetary importance as well as its programs.

Although there is no explicit plan for arms industries conversion, a de facto conversion exists, as can be seen in the big decreases in industry employment. Whereas the 1984–1988 Law of Programs (loi de programmation) had forecast increases in employment up to 345,000 direct jobs, employment in fact decreased throughout the 1980s to 251,000 in 1991. The various estimates (DGA, Unions) foresaw further reductions of about 50,000 more jobs by 1995. This phenomenon is common to most of the world's armaments companies, but France faces specific difficulties, particularly in its excessive level of exports - even more so because French exports are concentrated on a small number of products and countries.

The de facto conversion can be seen as well in the companies' industrial diversification strategies, which lead them to develop civilian production. This conversion varies according to the sectors concerned: it is quite advanced in the field of aeronautics, but not in land armaments. Moreover, changes in military strategy are creating new conditions for branches that were previously protected, such as the nuclear or shipbuilding branches. This conversion - which is a tactical adaptation to the new environment - has had varying consequences on the employment categories concerned; it has affected more workers than engineers and technologists, i.e., its effects are felt more on the production side than in research and development.

Military R&D is at the heart of strategic priorities in French defense doctrine. What is at stake is the ability to maintain a scientific and technological vigilance in order to maintain a technology base at the level of other countries. More fundamentally, various actors have moved towards virtual strategies in which the ability to perform is as important - or more so - than the performance itself. The French determination to preserve sovereignty - through autonomy in armaments - singles out military R&D as the sanctuary of autonomy and makes R&D conversion in the future difficult, even though economic constraints strongly weigh in that direction.

MILITARY R&D: ILL-DEFINED BUT COSTLY

Military R&D in France can be measured in different ways, but all the measures show that since the beginning of the 1980s the share of military R&D spending in public R&D financing has never been less than one-third of the total. This proportion increased in the latter part of the 1980s when civil R&D financing by the State rose more slowly, and the growth in total national expenditure for R&D was due essentially to the rise in military R&D spending.

Measuring military R&D (or 'defense research' as it is called in parliamentary documents) is difficult because it is closely determined by the needs laid down by the end-users, the military. This research is not therefore an autonomous activity that develops according to its own criteria, but an activity linked to the overall process of design, execution and use of the materials. Moreover, because of the costs of the arms programs and the length of their development stages (and therefore their economic impact), it has rarely been conceivable to defer manufacturing decisions to the end of the development stages: the irrevocable choice - the decision to launch the development - is made virtually at the beginning. R&D has thus become more difficult to pick out as a separate process and as a precisely identifiable budgetary ensemble. The integration of R&D into the overall execution of a program clouds any clear-cut identification of the R&D costs.

The concept most often used in France to measure defense research is the Research-Development-Tests (RDT) aggregate, which combines all appropriations under titles V and VI (capital expenditure) earmarked for research-development operations undertaken by the Ministry of Defence in the annual defense budget. This aggregate has changed over time as shown in Table 8.1. Although RDT may be of interest for comparisons over time, its significance is limited by a number of factors:

- part of the costs for personnel and operating expenditures of some departments of the Délégation générale pour l'armement (DGA) are inscribed in budgetary lines under the headings of these departments rather than under headings for research-development or studies;
- not all research and development expenditures inscribed under DGA appropriations are included, nor are those expenditures (equipment, operating) that are defined as being allocated to research and development financing, but whose execution is done outside DGA (as in the case of the Army health service, the Navy hydrographic and oceanographic service, or the Army technical services).

In order to remedy these shortcomings the military research and development budgetary expenditure (MRDBE) aggregate was drawn up in 1981 (see Table 8.1). This series provides a better measure of the volume of research and development spending in relation to overall defense expenditure, even if it cannot be used as such for international comparisons because of some differences in definition.[1] MRDBE takes into account the overall activity of research and development of the Ministry of Defence by an ex post analysis of expenditure allocations, regardless of their budgetary line entry.

Table **8.1**
Military Research and Development: Spending, 1981–1991

Year	RDT Aggregate (outlays in billions of francs)	MRDBE	Share of MRDBE in the defense budget
1981	13.26	15.67	15%
1982	15.47	17.30	14.1%
1983	15.22	18.16	13.6%
1984	17.54	20.24	14.2%
1985	18.25	22.37	14.9%
1986	19.17	24.46	15.4%
1987	21.98	26.62	15.7%
1988	26.11	32.41	18.6%
1989	27.80	32.98	18.1%
1990	31.34	34.69	19.5%
1991	30.36	33.00	18.5%

Sources: *Draft Finance Act, Rapport annexe sur l'état de la recherche et du développement technologique*, Imprimerie national Paris (annual); *Rapport annexe sur l'état de la recherche et du développement technologique*, various years; and *Armees d'Aujourd'hui, La défense en chiffres*, supplément annuel, February issue, various years.

The data leave no room for doubt: since the mid-eighties in France, there has been a rise in the share of military R&D expenditures in the total military budget, even while the overall pressure on the military budget has intensified and its share has fallen, both in the State budget and vis-à-vis the commercial GDP, from 4.01 per cent of the commercial GDP in 1983 to 3.26 per cent in 1992. This resistance of military research and development expenditure to the general downward movement of military spending clearly poses a serious problem for conversion, which is dealt with in detail below.

These data, however, cannot be used for international comparisons owing to differences in definitions; MRDBE, for example, includes VAT. Moreover, according to the Ministry of Research, 'corrections must be made in order to obtain for all research and development expenditures figures that are consistent with international survey data and criteria,'[2] although the Ministry does not specify the nature of these corrections. On the basis of these modifications, the Ministry has estimated that by international standards, costs for military R&D for 1988 were about 25 billion francs (compared with 32.4 billion francs for MRDBE) - a figure close to that published by OECD. OECD evaluations put the total French military research and development expenditure in second place in absolute value among OECD countries (after the United States, of course), but also - and this is significant - in first place in the share that military research and development represents with respect to defense expenditure.[3] This highlights even more the polarization of French military spending toward research. The place of these expenditures in the French strategy is pivotal. It is really a matter of acquiring or preserving, through a strategy of technological vigilance and the accumulation of knowledge, a power base for strategic autonomy, which is no longer presented primarily as a priority assigned to manufacturing but rather as shifted to R&D.

These observations on the pivotal role of military R&D in France can be qualified, since the share of military R&D in central government spending on all R&D is less in France (37 per cent) than in the United Kingdom (45.5 per cent), and is much less than in the United States (65.4 per cent).[4] But in the final analysis this distribution is not explained by less militarization of R&D in France, but rather by the fact that in Anglo-Saxon countries the tradition of state funding of civilian research is less powerful than in France, where the state's role is notably more important than that of private companies. This is not the case in the United Kingdom or the United States.

In the case of France, there is a clear priority in favor of military R&D. This choice is reflected in increased appropriations for it and in the general approach of the latest draft programming act. Appropriation restrictions imposed by economic stringencies[5] and changes in international relations are imposed primarily on the other items of the defense budget, such as ordinary expenses, manufactures, and stockpiles, not on R&D.

STATE OF RECONVERSION IN THE FRENCH ARMAMENT INDUSTRY

While there may be no explicit policy for conversion of the French arms industry, it is nonetheless clear from employment figures and other indicators that the scope of the contraction of the weapons sector in France is such as to amount to de facto conversion. According to Ministry of Defense figures given each year in reports to parliament, direct employment in the weapons industry represented 310,000 persons in 1982 and, following steady decreases, fell to only 251,000 persons in 1991. This 20 per cent drop was greater than the drop in total industrial employment over the same period,[6] and is the opposite of what was officially forecast in 1983, during the preparation of the 1984–1988 programming act, when the rapporteur specified that the volume of expenditure for equipment forecast would add 35,000 jobs to the 310,000 armament workers.[7] Some time, however, was needed for the reality of this declining movement to sink in, for it not to be looked upon merely as an economic downturn or 'hiccup.' Even in 1992 the rapporteur to the National Assembly on the arms industry cautiously went no further than to write, 'National armament production: a lasting decrease *probably*' (emphasis supplied).[8] Defence Secretary of State Jacques Mellick in an address to the Centre des Hautes Études de l'Armement (CHEAR) was more forthright: 'Let's not fool ourselves: there will not be, at least on the horizon of the end of the century, an increase in budgetary orders that might be expected were the present situation only a bad patch.'[9] Details of this general downturn in employment in the major companies in the sector are given in Table 8.2.

The general contraction has continued with further staff reductions announced by a number of companies. Comprehensive estimates published by the Conseil des industries de défense (CIDEF) paint a very bleak picture, with about 100,000 jobs threatened in the arms industries, out of a total of 283,000 canvassed by this institution (see Table 8.3). True, these estimates might have been given as a warning, coming as they did during deliberations on the 1992–1994 military programming act. Nevertheless, other estimates, be they from DGA or from economists, set the number of expected job losses in the industry at around 50,000, affecting a quarter to one-third of jobs in the arms industry by 1995. It is fair to say that with this latest development, coupled with the downsizing previously recorded, there has actually been a de facto conversion of a substantial part of the defense activities in France since the mid-eighties.

A Creeping Conversion

While this shrinkage of jobs in the French arms industry can be looked upon as a conversion, it must be said that it is a 'creeping' conversion; that is, unintentional. The word 'conversion' until just a short while ago did not form part of the jargon of French political leaders or arms makers; indeed, it hardly went beyond the circle of peace activists. Even the most daring persons in official or industrial quarters would go no further than to speak of diversification. Reflecting a change in attitude, however, the notion of conversion is no longer taboo, and is even making some headway in policy discussions in industrial seminars. But this evolution does not mean that conversion is now favored for its political and social values. It is only because the judgment on the French armament situation has changed; it now appears impossible to resist contraction. Under such circumstances it is more effective to go along with the movement than to buck it.

Table **8.2**
French Arms Companies: Employment 1982, 1991 (parent company)

Company	Personnel 1982	Personnel 1991
Thomson-CSF	40,148	22,250
Aérospatiale	36,450	26,748[10]
GIAT Industries	17,000	11,074
Dassault Aviation	15,782	11,914
SNECMA	12,595	13,816
Sagem	7,787	6,392[11]
S.N.P.E.	6,843	5,653
Matra	5,933	3,300[12]
Turbomeca	4,351	3,944
Dassault Electronique	3,238	3,416[13]

Source: Company annual reports.

Note: staff of non-consolidated companies. For other points of comparison, see J-P Hébert, 'France: ajuster le poids des armes,' *Economie et Humanisme*, no. 316 (January-March 1991).

The French arms industry sees a number of factors putting pressure on it; some are peculiar to the French situation, whereas others are general. To begin with there was a high degree of excess capacity in the arms sector: 40 per cent of production was exported (and thus 40 per cent of jobs were tied to exports). This excessive development was even more remarkable as companies in this sector had particularly heavy debt burdens and financial expenses.[14] It was therefore to be expected that a turnabout in the world market would have particularly marked consequences on the French arms industry, especially since its exports were highly concentrated in a small number of products (essentially aeronautic) and a small number of client countries (essentially in the Gulf region). These difficulties were exacerbated by the uncompetitive production conditions of certain establishments such as the land arsenals, before their reorganization and incorporation as GIAT Industries. The other factors are well known: the slump in world oil prices, limiting the resources of the traditional country clients; the shrinkage of arms expenditures of many Third World countries (except for the Indian subcontinent area and the Far East); the rapid evolution of the international situation since the 1987 INF treaty on the elimination of Euromissiles; the fall of the Berlin wall and the end of the American-Soviet confrontation in Europe; the drop in world military budgets and the ever-increasing costs of arms programs. All these factors have forced a re-thinking of production conditions and specifications.

Table **8.3**
Employment Forecasts (1992–1994) in French Defense Industries

	GIFAS (aero-nautic)	SPER (elect-ronic)	GICAT (land/ground)	Consolidated Total[a]
	Employment 1991			
Direct employees	118,700	50,700	50,000	194,000
Indirect employees	189,000	40,000	75,000	260,000
Total employees	298,700	90,700	125,000	454,000
Defense employees	140,000	57,000	125,000	283,000
Reductions in Employment Forecast for 1992–1994				
Direct employees	30,000-45,000	13,000	15,000	49,500
Indirect employees	70,000	11,500	10,000	80,000
Total employees	100,000-115,000	24,500	25,000	129,500
Defense employees	60,000	21,000	25,000	98,000

[a] Some firms belong to more than one syndicate.
Source: CIDEF estimates.

The final result is the substantial drop in armament jobs in France, hand-in-hand with a clearly expressed intention by company managers to diversify production so as to be less dependent on military production. But this 'conversion' is not something willed in advance, chosen for political or strategic motives. In France it is the pressure of the international market and the world situation which are the decisive factors, rather than domestic political choices. Conversion in France is thus a 'tactical' adaptation to the terrain, not a strategic approach.[15] It is far from being a permanent change; any turnabout in the world market would probably bring a resumption of French armament activity, as can be seen by the hectic scramble for arms in the Far East, where the 'big' markets are still to be found. The conversion was rushed, under the pressure of economic events, and not in any prepared or planned manner - hence the extent of the social consequences in terms of job losses.

So far this conversion has come about by accepting restrictions on what seems to be secondary from the perspective of French strategy, that is, on exports, while seeking to maintain the essential, that is, a capacity for national production for the major weapons programs. The 250,000 'armament' jobs (DGA figures) will probably be 160,000 around the year 2000. This is probably the minimum size for a production apparatus that would satisfy the objective of industrial independence, an independence that is necessary for political sovereignty and has so far been at the heart of French choices in defense.[16] It is these strategic and political variables which, in the last analysis, determine the evolution of the military industry and which have produced a 'de facto' conversion that differentiates between fields of production, touching manufacturing and R&D in different ways.

A Differentiated Conversion

The nuclear sector, for obvious strategic reasons, is a case apart. Up to now it has been more or less spared from any drop in personnel. A large proportion is employed by the délégation générale pour l'armement (DGA) and by the Commissioner for Atomic Energy (CEA), whose decisions on employment depend more upon overall policy than upon industrial or commercial considerations. The only perceivable effect so far has been the loss of 580 jobs at an Aérospatiale production site due to the decisions not to produce the S 45 missile (intended for the Albion plateau) and to freeze production of the Hadès missile. However, the draft 1992-1994 programming bill marked the start of a shift in this field.[17] Nuclear equipment - appropriations for which have been declining (in current amounts) since 1990 - will no longer occupy the sacrosanct place it has held for thirty years; the economic clout of this sector should diminish unless the decline in nuclear production is replaced by nuclear disarmament activities (neutralization, destruction), particularly in the new republics from the former Soviet Union.

Ship-building in France is essentially in the hands of the Direction de Constructions Navales (DCN). Its 31,000 work force has varied little since the end of the 1970s. Unlike the ground arsenals, which have been privatized, navy arsenals have kept their status and work under direct State control. This stability is explained by the fact that DCN makes nuclear ships, which are high priority systems. This stability is further explained by the fact that these ships require considerable maintenance and upkeep facilities, which must be permanently on hand, even if during repairs or renovation they are only partially employed. Employment in DCN has been relatively protected, since a substantial part of its production had been farmed out to civilian shipyards, and has been partly repatriated in order to compensate for the stretching out of orders, thereby transferring the burden of the slowdown to the commercial shipyards. But even major French nuclear programs have been affected by budgetary squeezes: the seventh and eighth nuclear attack submarines will not be built. Construction of the new generation nuclear submarine, the SNLE 'le triomphant,' has been slowed down. The building program of the Charles de Gaulle nuclear aircraft carrier has been held back by several months. The order for a second nuclear aircraft carrier, which originally was scheduled for 1990, has been postponed until after 1997. The very structures of DCN are changing; for example, by setting up DCN-International to carry out market research and by the separation of State and industry tasks in the organization chart. These choices are a prelude to a 'de-arsenalization' comparable to the transfer of the land arsenals to GIAT Industries.

The aeronautical situation is more complex because of overlapping civilian and military activities within the enterprises. French aeronautics has increased its employment slightly over ten years by about 5,000 persons to about 120,000 persons, but only by altering the production mix. Whereas ten years ago less than 30 per cent of its turnover was for civilian production, today this share has almost doubled. These results are attributable to the success of Airbus, regional transport planes, and CFM 56 engines. The situation, however, differs from company to company: Dassault Aviation, very much involved in military aeronautics, has been hit much harder than Aérospatiale. But it must be noted that the goal of all companies is to increase the share of civilian production. In Dassault Aviation, military production was 88 per cent in 1980; in 1990 this share fell to 72 per cent. The announced target was to reduce this proportion to 55 per cent in 1995. Similarly, Aérospatiale, whose sales were two-thirds military in 1985, expected to lower this proportion to 40 per cent in 1995. At SNECMA, military sales accounted for 80 per cent of turnover in 1980, but only 36 per cent in 1990.[18] It is certainly in the aeronautics sector that conversion is best established: companies have sought to diversify their activities, to promote civilian production, and possibly to cut down by dropping certain manufacturing activities in favor of developing research activities.

In defense electronics, whose share of armament programs is steadily increasing, the situation is different. This sector is dominated in France by Thomson-CSF, which in ten years has become France's top arms producer. Contrary to the direction taken by the aeronautic industry, Thomson-CSF's focus is on 'professional' (read: military) electronics. It is a risky strategy if the downturn in production hits the electronic industry as well. But Thomson-CSF, a public company, is confirmed as 'the' national producer; it is the most important company in this sector, and so important that the State cannot abandon it. Although there have been sizeable personnel variations in Thomson-CSF, generally downward, it is very difficult to analyze them precisely in terms of conversion because of the frequent changes in the Thomson-CSF organization chart. Other companies specializing in electronics (Sextant Avionique, SFIM, SAGEM) are faced with serious problems, and they are obliged to envisage important civilian diversification.

It is in ground weapons that the need for conversion is the most pressing because of the general levelling-off of export markets and the decrease in size of national programs, which are linked, inter alia, to price movements, the vigor of both traditional and new competition, and to technological developments in arms systems whereby mechanics is less and less important. Some producers - e.g., Luchaire and Mnurhin - have vanished or been taken over by GIAT Industries. GIAT Industries, which has already slashed staff, is conducting a policy of both imposing itself as an unavoidable European producer, e.g., through its buy-out of FN-Herstal and PRB, and of developing its civilian activities. GIAT Industries' goal is to employ about 2,000 persons out of a little over 12,000 total employees in its civilian production by 1995 by setting up a Gitech division within the company which will bring together all current or potential civilian capacities of GIAT Industries.

Decrease of Production, Stability of R&D

These developments have affected the aerospace industry's production side more than its research sector, as illustrated by the data in Table 8.4. The same development can be seen in electronics; there has been a clear drop in the proportion of production workers, while the proportion of engineers in total employment has risen sharply (see Table 8.5). Depending on the sector, the future prospects for the French arms industry therefore vary greatly, but the severe effects of job cutbacks weigh upon all these sectors. That is why, in August of 1991, the

Ministry of Defense appointed an official to be in charge of restructuring. The 'Delegation aux restructurations' had a first-year budget of 100 million francs, which was increased to 240 million francs for 1994. That is a beginning. We can observe that the public authorities begin to understand the necessity of conversion, but only the conversion of excess capacity. In any event, it is clear that regardless of the forms this movement may take in the future, military R&D will be, in the case of France, the last sector to be affected, because of the French view of strategic questions.

RESEARCH AND DEVELOPMENT IS AT THE HEART OF FRENCH ARMS POLICY

Research and development is, for the future, at the heart of French arms policy. Its strategic importance has grown over the past few years, and this can only continue. No defense policy can be defined without including the essential choices vis-à-vis R&D. There is a remarkable shift: whereas the François Fillon report which introduced the 1987–1991 draft programming bill could still refrain from presenting an overall analysis of R&D policy, the 1992–1994 draft programming bill placed R&D at the heart of choices to be made, justifying priorities proposed for space, education, and technological vigilance.

Obviously R&D is needed to maintain a scientific and technical edge; this aspect is more timely than ever. The goal of French leaders is not only to prevent any circumvention of the means of deterrence, but also to be able to anticipate scientific and technical developments that would raise questions about or modify the capacities of any given system. From this viewpoint, it is clearly essential to be able to follow the evolution of fields concerning all aspects (acoustic and otherwise) of monitoring submarines, the field of anti-ballistic techniques, and electronic countermeasures. But it is equally important not to overlook conventional armaments, for example, future developments in anti-helicopter defense and explosives.

The challenge for the future goes beyond 'keeping up'; it entails adding a new dimension to the French perspective. Clout in international negotiations will depend more and more on R&D capacity. Investment in R&D is the price to pay for having one's voice really heard in these negotiations: in the competition for military effectiveness, relationships of strength depend more and more on potential – not yet produced – weapons, on results of research, or even on research projects. The

United States' SDI program is the most striking demonstration. R&D furnishes these potential means of influence.

Table **8.4**
Data for Personnel Movements in Aeronautics
(% of employment by type of activity)

Year	Studies	Prototypes	Production
1986	23.3	5.3	71.4
1987	28.2	5.5	66.4
1988	28.9	6.5	64.7
1989	28.4	6.3	65.3
1990	31.6	5.4	62.9
1991	32.3	5.3	62.4
1992	33.2	5.4	61.4
1993	33.9	4.8	61.3
1994	33.8	4.1	62.1
1995	33.6	3.4	63.0

Table **8.5**
Data for Personnel Movements in Electronics
(% of employment by type of status and total employment)

Year	Engineers	Technicians	Production Workers	Total Employment
1955	11	41	48	23,500
1960	11	45	44	32,000
1965	14	46	40	34,500
1970	16	47	37	38,000
1975	18	51	35	44,700
1980	19	47	30	47,000
1985	21	53	26	55,000
1990	29	50	21	50,900
1995	42	44	14	41,400

Moreover, French strategy cannot overlook the current movement, owing to the pressure of public opinion and policy shifts by the leading powers, towards more negotiations on armaments than in the past. France is not on the sidelines in these negotiations. France's weight in these negotiations, however, will depend on its capacity in 'potential means' added to its existing ones, not only to be heard as an independent voice, but also to affect the contents and objectives of these arrangements. The 'capacity to act' brings the capacity to 'renounce acting,' which is increasingly becoming a bargaining chip in international relations. By having these bargaining chips France will be able to avoid international agreements that disproportionately limit its own strategic means.

Negotiating power may be acquired through R&D, but this way is not free of uncertainty. It could gradually lead to more and more participation in the race for performance and a drift towards a strategy of attempting to outclass others technologically, as can be seen by current parliamentary reports after the Gulf War. In French doctrine, R&D must provide the negotiating means while avoiding participation (necessarily speeded up) in the race for sophisticated equipment.

At present, the cost of research and development accounts for about 30 per cent of all arms production. This is a heavy burden, and it is increasing. This constraint explains why political and industrial leaders are beginning to explore certain ways of bringing down costs, such as dual use techniques, i.e., changing certain military specifications to make greater use of civilian production. It was also with a view to lowering costs that the Independent European Programme Group (IEPG) launched the joint program of military technological research, 'Euclid.'

Moreover, the tempo of technical progress introduces a fresh constraint: studies started in 1990 will give rise in the year 2000 to the manufacture of weapons that will still be in service in 2030. These developments will be less and less definitive systems and large-scale programs, and more and more systems that must be adaptable, modular, flexible in design and suited for mutual exchanges with civilian production. The thrust of all this scientific and technical movement is to bring French arms policy more and more to a strategy of planned upgrades to successive generations of weapons, a strategy of means, and even of potential means.

CONCLUSION

Conversion of military R&D in France is a very touchy matter because of its very strong ties to the State. While de facto conversion may be seen in military production, in R&D there is no true conversion or even a hint of such a movement. The only movements are limited to a shift in the field of nuclear research, in keeping with the evolution of the strategic doctrine where the role of nuclear weapons is being re-examined.

NOTES

1. Fuller data for a more precise picture of the methodological problems involved in the definition of military R&D expenditure can be found in: J-P Hébert, *Stratégie française et industrie d'armement* (Paris: FEDN, 1991), ch. 6. See also F. Chesnais and C. Serfati, *L'armement en France: Genèse, ampleur et coût d'une industrie* (Paris: Nathan, 1992), pp. 107–11.
2. *Draft Finance Act, Rapport annexe sur l'état de la recherche et du dévéloppement technologique* (Imprimerie nationale, Paris, 1991), p. 18.
3. Cf. Tables 1.1 and 1.2, Chapter 1, this volume.
4. OECD, 'OECD en chiffres: statistiques sur les pays membres,' *Supplement à l'Observateur de l'OCDE*, No. 170 (Juin-Juillet 1991), p. 54.
5. See also J. Fontanel, and J-P Hébert, 'L'importance économique de l'industrie française d'armement,' in 'Economie de la défense,' ed. J. Fontanel, *ARES* 13, no. 4 (1992), p. 5.
6. See J-P Hébert, *Stratégie française et industrie d'armament* (Paris: FEDN, 1991), chapter 7.
7. See L. Tinseau, *Rapport sur le projet de programmation militaire 1984–1988*, Assemblée nationale, 11 mai 1983, Doc. No 1485, p. 73.
8. J-G Branger, 'Avis/../sur le projet de loi de finances 1993,' *Recherche et industrie d'armement*, Assemblée nationale, Document 2948, Tome VI (14 octobre 1992), p. 10.
9. Branger, 'Avis/../sur le projet de loi de finances 1993,' p. 21.
10. The sharp drop in personnel in 1991 was due to company affiliation of 'helicopter' activities within Eurocopter.
11. 1990 data.
12. Data (1990) not strictly comparable due to affiliations of defense activities.
13. Personnel in Dassault Electronique peaked at 4198 persons in 1987.
14. See J-P Hébert, 'L'effort militaire française et ses retombées sur l'économie,' in 'Economie de la défense,' ed. J. Fontanel and J. Aben, *ARES* 12, no. 4 (1990), pp. 47–70.
15. See J-P Hébert, 'Between Disarmament and Sovereignty: The Issue of the Conversion of Arms Industries. The French Example' in *Conversion: Opportunities for Development and Environment*, ed. A. Brunn, L. Baehr and H-J Karpe (Berlin: Springer-Verlag, 1992).

16. In 1983, when officially no one expected such a cutback in the sector, a weapons engineer wrote 'at three hundred thousand, we manage somehow or other, but at two hundred thousand we must think of closing up shop.' F. Varenne, 'Ventes d'armes: le juridisme et l'incantation,' Projet, No. 177.
17. See J-P Hébert, 'Pour comprendre la nouvelle loi de programmation militaire 92–94,' *Alternatives non-violence*, No. 85 (Winter 1992).
18. Company annual reports, various years.

9

Dual-Use and Conversion of Military-Related R&D in Germany

Jürgen Altmann, Wolfgang Liebert, Götz Neuneck, and Jürgen Scheffran

INTRODUCTION

For decades, both West and East Germany had to carry heavy burdens in terms of military forces, maneuvers and deployment of foreign troops close to the iron curtain. Germany was one of the most densely militarized zones in the world; the inner-German border between East and West created tremendous risks and burdens including the threat of extinction in a nuclear war.[1] On the other hand, after 1945 Germany never possessed a highly militarized industrial base as did the United States, although several sectors were heavily dependent on military contracts (e.g., ship-building, aerospace, machine-building). In the shadow of the superpower competition and within the European Community, West Germany was able to rebuild its economy and gain economic competitiveness. In this respect, Germany, like Japan, profited from the Cold War. With the fall of the Berlin wall a new situation has arisen, in many respects the reverse of the East-West conflict. Most of the security risks of a European military confrontation have vanished, but economic problems have been aggravated due to unification.

Germany does not have a comprehensive conversion program, but at least in theory a broad understanding of the concept of conversion after the Cold War can be found among experts:

> Conversion covers not only the shrinking of military forces due to budget constraints after the end of the Cold War, but also the

> transformation of finances, the modification of research and development (R&D), the reintegration of soldiers, the alternative use of barracks, land property and military infrastructure including the civilian use and the dismantlement and destruction of military equipment.[2]

In the past the conversion debate in Germany concentrated largely on economic and social aspects of converting military production facilities and integrating them into the civilian economy.[3] Little attention was given to the R&D complex which forms the early stages of the weapons life-cycle. But an effective disarmament strategy should also include the conversion of military science and technology. Preventive arms control is also needed to make conversion irreversible and to avoid dangerous military-technological arms races and emerging proliferation problems.[4] In the longer run Germany might profit from reduced military spending by converting its military infrastructure, including the military R&D complex.[5] However, as in other Western countries, theory often does not meet reality: conversion is more difficult than originally assumed. Besides the practical problems of transforming the arms industry into a competitive, market-oriented civilian industry, the biggest obstacle is renewed interest in the development, production and export of arms as part of an extended security concept.

THE NEW GERMAN SECURITY CONTEXT

At present no specific military threat can be identified for German security. The victorious powers of World War II are removing their troops from German territory.[6] The new Germany is now surrounded by friends and partners seeking economic cooperation; it is integrated into the new European security architecture which includes NATO, the Western European Union (WEU), the European Union, the Conference on Security and Co-operation in Europe (CSCE) and the European Council. Within the CSCE, confidence building measures and arms control agreements with comprehensive verification mechanisms improve military transparency in Europe and encourage disarmament. Since unification the German unified military forces have been reduced from about 650,000 to 420,000 soldiers and they are to be further reduced to 370,000 by the end of 1995. The Treaty on Conventional Forces in Europe (CFE) requires the reduction of more than 10,000 Treaty Limited Equipment (TLE) by the German armed forces.

Germany is an active member of NATO, the most powerful military alliance in the world. Whereas during the Cold War NATO opposed the Warsaw Treaty Organization, now the 'Partnership for Peace' aims at political integration with East-European countries. Based on a diffuse threat perception of 'new risks, uncertainty and instability,' the new NATO strategy aims at creating mobile force structures, which are to be implemented in the member countries to guarantee the capability for counter-offensive and counter-concentration in every point of the defense area. Earlier debates in Germany on a non-offensive force structure and the orientation toward a defensive posture seem to be forgotten.

Without a clarification of the future tasks of the Bundeswehr, long-term budget planning and re-organization of military R&D is difficult. For Germany, the central security task remains the defense of the country and the alliance, but all options for an 'extended defense' outside of the NATO territory are being kept open. The basic goals of security policy and the mission of the Bundeswehr have been defined in the 'Defense Policy Guidelines' of the Ministry of Defense (Bundesministerium der Verteidigung, BMVg).[7] According to these guidelines, the 'vital security interests of German policy' include not only the protection of Germany, strong ties to the alliance, European integration and crisis management, but also the 'continuation of free trade and unimpeded access to markets and resources worldwide as part of a fair world economic order.' Similar objectives have been expressed in U.S. strategy plans.

The constitutional court decided in July 1994 that 'out-of-area' operations of German military forces would be allowed within the framework of a collective security system. Thus, the participation in peacekeeping operations outside the country is possible. Before sending troops, the German Parliament has to approve the action by simple majority. Of course, this will have wide-ranging consequences for future procurement and the on-going doctrinal debate in Germany, with a direct impact on future military-related R&D.

Far-reaching structural changes can be observed for the military forces. A number of programs serve the increase in mobility and a fast relocation capability over wide areas. In the army the material planning is directed towards a 'light army of the future,' with air mobility, air relocation and air mechanization playing an increasingly important role in army planning. To achieve this goal improved C^3I capability is required. Various guided weapons and submunitions are in development. Part of the equipment is to be procured for UN actions and, in the peacekeeping context, a military space program is under consideration by the government. Finally, there is an interest in Europe in pursuing integrated air and missile defense systems, based on ten years of R&D in the U.S. Strategic Defense Inititative (SDI).

THE GERMAN ARMS INDUSTRY AFTER THE COLD WAR

Germany's arms industry was built up during the 1950s and 1960s, and is capable of developing and producing a wide range of major weapon systems.[8] In particular, Germany has well developed nuclear energy, aerospace and ship-building industries. The arms industry is integrated into various industrial branches, which produce civilian goods as well. Due to the changed security context, it is facing hard times. The burdens of unification, the economic crisis, reductions in investment, reduced defense budgets, export regulations, and attitudes in the public hostile to excessive military activities all lead to the conclusion that the existing military and industrial infrastructure is oversized.

A few figures describe the current situation:[9]

- According to official numbers,[10] the defense budget increased nominally from 42.6 billion DM in 1981 to 53.6 billion DM in 1991 before falling to 50.15 billion DM in 1993, which in real terms corresponds to the defense budget of 1986. The defense fraction of the federal budget shrank from 18.3 to 11.5 per cent between 1981 and 1993 and its fraction of the GNP from 2.77 per cent to 1.57 per cent.
- Substantial cost savings of 5.6 billion DM annually are expected from further reductions in the number of employed soldiers and civilian personnel. The number of civilian employees in the BMVg will decrease by about 40,000 by the year 2000. In the same period, half of the 102,000 civilian employees working for the allied forces, could lose their job.[11] The number of military sites in West Germany has been reduced from 720 to 589. However, the savings were not - as planned - directed to defense investments but were mostly used for the consolidation of the federal budget.
- Due to the budget constraints, the restructuring of the European arms industry, and the decline of arms exports, the number of employees in the German arms industry fell from 280,000 in 1989 (including 40,000 in the former GDR) to 180,000 in 1993.[12]

Three major strategies to deal with the current problems can be distinguished: deep reductions in the military arsenals and conversion of the defense industries; a restructuring combined with a qualitative modernization of the military arsenals by pursuing development and production of high-technology weapons at a lower quantitative level, applying the dual-use strategy (see below); and internationalization of arms production and development, and exports of armament and related technologies, especially to developing countries.

At the level of the firm, some companies are selling their plants and subsidiaries that specialize in armaments and withdrawing completely from the field. Others are diversifying in order to diminish their dependence on military orders, or they are dismissing employees. Larger enterprises increasingly are merging in international consortia in order to raise their share of a decreasing financial budget.[13] The future of Germany's arms industry is seen to lie in the internationalization of the Western arms industry, in particular, in the collaboration with other European countries and the United States within NATO on new weapon technologies. In the framework of the Maastricht Treaty the path to a 'European Armament Agency' is pursued with some WEU and NATO member states.[14]

The major reasons for international cooperation are reduced budgets, a reduction of procurement numbers, and high R&D costs, which together lead to rapidly increasing costs per unit, and a reduction in demand. Declining demand leads to further dismantling of capacities and a loss of know-how.[15] European cooperation is an attempt to share R&D costs and create a bigger (European) market for fewer products. In principle the consequences of such cooperation include an international division of labor, standardization and specialization in military capabilities, but, according to defense planners, the existing capacities in the German arms industry are to be protected, in particular a core workforce and production capacities in the tank industry, munitions industry, aircraft and electronic branches.[16]

The state of the German arms industry is exemplified by the air and space industry, which traditionally has a high military share. After a peak in 1989/90, the civil aviation business experienced a sharp decline. Because it was formerly dominated by government demand, the aerospace industry is highly dependent on the financial situation of the state. Compared with the United States or France, there is less consensus in the German society on the need to maintain the aerospace industrial base.

According to the Bundesverband der Deutschen Luftfahrt-, Raumfahrt- und Ausrüstungsindustrie, BDLI (Federal Union of the Aerospace and Equipment Industry) the number of employees fell from 95,000 in 1990 to 86,300 persons in 1991, who nevertheless produced an increased turnover of 26.77 billion DM. This trend continued in 1992. In the medium term, a fall in the level of employment by 20 per cent is expected, with an increasing civilian share.[17] Spaceflight has stayed constant with 6,500 employees. In the engine industry the number of employees fell by 6 per cent from 9,250 (1990) to 8,590 (1991), with the military fraction sinking by 18.3 per cent. In the equipment industry, with a military share of 55 per cent, the reductions were especially large. Although the number of employees and orders for aerospace defense were reduced (the latter by 14 per cent), an increased turnover was still produced.[18] A number of military aviation programs are being continued: the modernization of the F-4F PHANTOM, the experimental aircraft X-31, and FANRANGER; and the delayed development of the EFA 2000, the Tornado Emitter Locator System, the NATO helicopter NH-90 and the PAH-1 anti-tank helicopter. For spaceflight the future is open, with the focus on communications and earth reconnaissance satellites.

The single most important event in the German military and aerospace industry was the 1991 takeover by Daimler Benz of the companies Telefunken, Dornier, MBB and MTU, which were merged to form the Deutsche Aeropace AG (DASA). MBB was restructured in May 1992; in July, DASA and AEG founded TEMIC to fuze their microelectronic activities. The arms sales of Daimler-Benz are more than four times as high ($3920 million 1991) as those of the second largest company, Siemens ($900 million), and third, Rheinmetall ($860 million).[19]

European cooperation is increasingly important in military aerospace, especially between Germany and France (e.g., the Eurocopter International venture formed by Aérospatiale and MBB in December 1991). Examples of cooperation are the missiles Stinger, Roland, Patriot, Milan, HOT, PARS 3, POLYPHEM and Apache. Cooperation also extends to equipment. In all these programs, extensive R&D is required. In December 1993 the French and German Defense Ministers F. Léotard and V. Rühe announced the creation of a Franco-German Armaments Agency, which has the task of managing the bilateral armament programs more efficiently.

German policy in arms trade is contradictory. On the one hand, Germany has implemented new far-reaching reforms of export controls on dual-use goods, but on the other, it has increased its direct weapons exports in the early 1990s. According to SIPRI, German arms exports (in 1990 prices) increased from $953 million in 1989 to $1.67 billion in 1990 and $2.41 billion in 1991, before leveling off at around $1.85 billion in 1992 and 1993. In 1992 Germany became the third-largest exporter of major conventional weapons and the largest in Europe, accounting for 41 per cent of deliveries by EC countries versus 24 per cent by France and 20 per cent by the UK.[20] According to the UN

Register of Conventional Arms, Germany was in 1992 the second most important supplier of the main combat weapon systems, with deliveries in each of the seven categories.[21] Most of the exports in 1992 were significant deliveries of second-hand equipment from the former Nationale Volksarmee (NVA) of the German Democratic Republic - mainly tanks and ships - to NATO countries or Finland and Sweden. It remains to be seen if this trend will continue.

CIVIL AND MILITARY RESEARCH AND DEVELOPMENT IN GERMANY

Military-related R&D plays an important role in implementing new strategic concepts in most of the leading industrial countries. Following the end of the bipolar confrontation, the military planning philosophy in Germany as in the United States is to do a lot of research and development, stretch procurement programs, and not build everything immediately.[22] The new 'White Paper 1994' of the BMVg states: 'By concentrating R&D on technological key areas the option is maintained to pursue the required modernization later on.'[23] The German Minister of Defense Volker Rühe said in a speech at the German 'Führungsakademie' (officer's academy) in Hamburg in 1992:

> In the future we need less and - in a wide range of fields - different material . . . For that purpose we have to reverse the situation in quantitative and qualitative terms . . . we need technology of the highest quality and performance as a 'Force Multiplier.'[24]

For this purpose, civil technology will be increasingly integrated into military development as add-on programs.

The German Science and Technology Budget and Structure

In many fields of research and technology, Germany is among the leading nations in the world. In its share of the gross domestic product (GDP), Germany's R&D expenses in 1990 (2.77 per cent) were behind Japan (3.07 per cent), equal to the United States, but ahead of France (2.44 per cent) and Great Britain (2.21 per cent).[25] In 1991, the German national R&D expenses amounted to 76 billion DM, corresponding to 2.66 per cent of the GDP. This decline in the R&D share in GDP, which began in 1989, continued in 1992. The largest fraction of R&D expenditures in

1991 (59 per cent) was paid by the industry, followed by the federal government (22 per cent). 1993 was the first year that the federal R&D budget fell in nominal numbers, from 17.97 billion DM in 1992 to 17.94 billion DM in 1993 according to the government plan.[26]

The agency mainly responsible for research and technology was the 'Bundesministerium für Forschung und Technologie,' BMFT (Federal Ministry for Research and Technology); which in 1995 was fused with the Ministry of Education and Science. Its 1993 budget was 9.4 billion DM, more than half of the 17.9 billion DM that the German Federal government spent for R&D that year.[27] Other principal German government funders of R&D are the Defense Ministry (BMVg), the Economics Ministry (BMWi), and the Ministry of Education and Science (BMBW).

The BMFT is the main funder for Germany's four major publicly funded research institutions, and its priorities and costs are reflected in the research areas they cover. The institutions are:

- The sixteen Grossforschungseinrichtungen (GFEs), or large research organizations, working in a variety of fields from energy to advanced materials, information technology, environment, aeronautics and space.
- The Max Planck Society (Max-Planck-Gesellschaft, or MPG), which performs basic scientific research, mostly in the natural sciences.
- The Fraunhofer Society (Fraunhofer-Gesellschaft, or FhG), the smallest of the four major research institutions, whose task is to pursue applied R&D dedicated to direct transfer into practical use.[28]
- The Blue List, which is not an organization, but a collection of independent research organizations, including a number of new East German institutes, working in various fields.

Military R&D

Since 1954 when rearmament began, military influence in the research and industry sector in Germany has never been as dominant as in the United States. Although military interests played a prominent role for politicians like the former Minister of Defense and of Nuclear Affairs, Franz-Josef Strauβ, throughout the 1960s direct investment in civilian R&D was seen as preferable to indirect spin-offs from military R&D.[29] This attitude continued until the late 1980s when former U.S. President Ronald Reagan's Star Wars program had challenged the Europeans to focus more strongly on military high technology. After the end of the East-West conflict, military planners in Germany, as well as in the

United States, see qualitative improvements of weapon systems as a key element to compensate for the quantitative arms reductions that are expected in the coming years. Consequently, for some planners, R&D forms the basis for new and improved defense products, for the 'automatized battlefields' in particular, as well as 'out of area' conflicts requiring mobility and flexibility.

Expenditures for military R&D are usually taken to comprise all officially declared R&D in the defense budget plus the declared military R&D in other government budgets. Much more is known about the governmental budgets than the military-related spending in industry, although the latter perform the main part of the application-oriented R&D. Table 9.1 presents the official figures for military R&D expenditures for the period 1981–93. As a consequence of the continued interest in defense-related R&D, military R&D is shrinking more slowly than the total military budget. There has been a decline in the military share of federal R&D since 1990, from 22.3 per cent in 1990 to 18.2 per cent in 1992. Between 1981 and 1990, however, BMVg R&D had more than doubled in absolute numbers, and the 1992 share is still much higher than in 1981.

What is counted as military R&D? The official figures omit important categories of spending. At least for 1990 better figures are available.[30] The Federal government has admitted that the costs of running the universities belonging to the Bundeswehr and the technical and scientific offices of the Bundeswehr, funds for international research institutions, and staff costs within the R&D sector have to be added. In 1990, this amount was roughly 400 million DM.[31] Together with some smaller figures this brings the sum up to 4,010 million DM. Furthermore, the government reimburses the industry by between 3 to 6 per cent of the production costs of military goods that it purchases, under the headline 'additional for free research and development.' It is unclear how much was paid in 1990 in this way. But based on procurement figures one could estimate a number of about 500 to 1000 million DM. So it turns out that in 1990 the total amount of military R&D spending by the Federal Government was about 4.5 to 5 billion DM, compared to the official figure of 3.4 billion DM. Estimating the same proportion for 1993 one arrives at around 4.8 billion DM, versus the official figure of 3.2 billion DM. One can assume that the same correction would apply to every year, so that one can add roughly one third or one half to the official number for military R&D. The federal government has admitted that for 1990 the 'true' military R&D budget was at least 4.01 billion DM, but this figure has never been included in official publications.[32] It would correspond to a 16.1 per cent share of total governmental R&D

(26.4 per cent of the federal R&D budget), much higher than the official numbers transmitted to OECD.

Table **9.1**
R&D expenditures in Germany, 1981–93 (million DM)*

Year	Federal R&D	Gov. R&D	R&D of Business and Enterprise Sector	Federal Military R&D	Military Share of Gov. R&D	Military Share of Federal R&D
1981	10,371	17,261	22,082	1,528	8.9%	14.7%
1983	11,438	18,379	25,459	1,895	10.3%	16.6%
1987	13,152	21,694	36,831	2,894	13.3%	22.0%
1989	14,036	23,113	41,197	3,128	13.5%	22.3%
1990	15,149	24,780	42,500	3,389	13.7%	22.3%
1991	16,853	28,990	46,045	3,170	10.9%	18.8%
1992	17,969	30,780	47,520	3,263	10.6%	18.2%

* Official figures. Governmental R&D includes expenditures of the German federal government and the 16 state governments.
Source: BMFT, *Bundesbericht Forschung 1993*, pp. 547, 552-59; Bundestagsdrucksache 11/7373 (1992); E. Bulmahn (see footnote 30).

Military R&D Concepts

The priorities and objectives in military R&D are determined by the Research and Technology (R&T) Concept of the BMVg, which was introduced in 1985. Defense-related R&D is divided into three areas:[33]

1. Research oriented towards defense technology,
2. Future key technologies oriented towards military tasks,
3. Systems technology/guiding concepts.

Military tasks in research and future technologies include:

- reconnaissance, command and control (e.g., radar improvements, sensors, GaAs technology, optronics, pattern recognition, friend-foe-identification);

- mobility (e.g., light-weight construction, fluid-mechanical improvements, remotely piloted vehicles, energy support for electromagnetic guns);
- weapons use and lethality (e.g., electromagnetic guns, weapon lasers, torpedoes and torpedo defense, land mines, kinetic weapons against improved armor);
- protection (e.g., improved armor, ABC protection, protection against microwaves and electronic warfare);
- planning support (e.g., military-technological trends, threat perception, simulation technologies, computer-aided planning, international harmonization of armament programs).

The Fraunhofer-Institut für Naturwissenschaftlich-Technische Trendanalysen (INT, Institute for Scientific-Technical Trend Analyses) at Euskirchen near Bonn provides an evaluation of relevant literature on a regular basis for the BMVg, which is used as an input to the 'defense technology forecast.' In this forecast, long-term projections of trends in military technology are presented and evaluated by the BMVg. The 'White Paper 1994' states that 'In the Research and Technology Concept of the BMVg, priorities are determined according to military requirements, which include civilian Research and Technology projects of the government, international cooperation and industrial research results.'[34] For a deeper analysis, however, it seems almost impossible to link the official R&D objectives to concrete projects and funding in the federal budget.

Cooperation Within NATO and WEU

International cooperation in military-related R&D is seen as essential by the BMVg: 'The BMVg focus is on cooperation: 70 per cent of the large projects were realized in international cooperation.'[35] In June 1989 the NATO Independent European Programme Group (IEPG) established a military cooperation program, the European Co-operation for the Long Term in Defence (EUCLID), which consists of 11 so-called Common European Priority Areas or CEPAs (e.g., new radar technologies, artificial intelligence, stealth, etc.).[36] Since then the number of CEPAs has further increased. In the Program Memorandum of Understanding, signed at the end of 1990, 26 Research and Technology Projects (RTPs) were included to structure the CEPAs. At the end of 1990, 35 RTPs were under preparation in Germany with spending at 55 million DM. In 1992, the volume had increased to 72 million DM, with a projected annual increase of 20 million DM.[37] The ministers of the IEPG agreed in

1992 to transfer the functions of the IEPG to the WEU, and the IEPG is now known as the 'Western European Armaments Group' (WEAG). This marks a step towards European armament cooperation in terms of harmonising requirements for joint programs, research, and trade as a step towares developing a European defense market.

More specific information than this on German R&D projects and contractors is unavailable. Military R&D in Germany is still planned and performed under secrecy, and information on concrete projects is hardly available. Improved transparency on military R&D is a prerequisite for detailed consideration of conversion of related research projects and institutions.

THE DUAL-USE STRATEGY AND ITS IMPACT ON R&D CONVERSION

Scarce resources, technological dynamics, and political pressures, combined with the converging demand profiles of civilian and military technologies, all support the recourse to a 'dual-use strategy,' whereby a technology is developed first for the civilian sector and then used for military purposes. For some planners 'dual-use' is a magic word that keeps the paths for further military-technological developments open. In Western Europe, as well as in the United States, the dual-use strategy aims at combining military interests in cheap and effective armament technologies with commercial interests in the high-tech sector. Research programs like EUREKA, ESPRIT and EUCLID have been designed as dual use from the beginning.[38]

The specific character of Germany's dual-use strategy was already outlined in 1986 by the then Permanent Secretary in the BMFT, Hans-Hilger Haunschildt:

> Given the importance of modern technologies for military as well as for civilian applications, both domains of responsibility, the BMFT and the BMVg, are closely working together in some subfields . . . The dominant basic technologies (are) often identical, e.g. the fundamental physical technologies, the materials sciences, the machining technologies, the semiconductor technology or data processing . . . Especially Germany and Japan have successfully demonstrated after the war that civil technological developments do not need the detour via government-funded military research and development. The

> 'American way' is no alternative for us . . . The primary civilian way also works.[39]

The philosophy that civilian R&D should be planned for the fulfillment of military goals has been realized in various programs, in particular, in the space program (e.g., by using civil satellites for military reconnaissance and communication) and in the 1989 concept for future information technologies.[40] In 1990, the German government stated that dual-use is a basic ingredient of its research policy: 'The Minister of Defense bases as a matter of principle his science and technology programs on civilian science and technology. Only when unavoidable the civilian scientific-technological base is complemented by specific programs.'[41]

The German government has repeatedly announced its intention to include military requirements early within civilian R&D programs and then exploit the results in 'add-on programs' for military applications. This implies not only that converging civilian and military development paths are used, but also that civilian research will be adapted to military purposes. Based on this strategy a 'grey area' of civil-declared but military-used R&D could develop. An essential element of this strategy is closer cooperation between the BMFT and the BMVg:

> With the BMFT the dialogue about research and technology cooperation must be intensified and deepened. The R&T activities with the potential for dual-use applications must be coordinated more thoroughly in advance between both ministries and initiated jointly, if necessary. So-called add-on programs of the BMVg are only possible, in principle, if the BMFT or/and the industry have created the appropriate basis.[42]

How the coordination process between the ministries works in detail, which programs - besides those officially mentioned - are integrated into dual-use concepts, is not known. Only the existence of the concept itself is documented.

In the context of the dual-use policy, diversification, which is a strategy to mix civilian and military R&D, will be more probable than irreversible conversion in the next few years. In this way, military objectives could profit from both civilian and military R&D activities and, even more, the reshaping of powerful military R&D programs is always possible. Obvious impacts of this diversification strategy in the field of arms production and arms industry are:

- the persistent option for militarily useful outputs of diversified R&D;

- reversibility of 'conversion' activities;
- continuation of secrecy for parts of R&D;
- persistent distrust.

As a result of the dual-use strategy, civilian R&D is increasingly amalgamating with military R&D. The R&D work of the BMVg, which cannot fund basic research, is increasingly based on civilian resources, which in turn are also influenced by military purposes. Unlike the United States, where dual-use is closely connected with preventing deep cuts into military R&D, in Germany the dual-use strategy could result in expanding the influence of the military on some part of civilian R&D. The extended notion of military-related science and technology is, in general, contrary to further disarmament and conversion efforts, as it further blurs the differences between civilian and military R&D and allows the performance of military R&D under a civilian cover. The spin-offs from the military to the civilian sector have been disappointing; now the transfer into the opposite direction, or 'spin-on' is organized. Civilian R&D has become a priority for the military, which can use the outcomes for its own parallel military R&D. The question is to what degree the dual-use concept leads to a clandestine militarization of R&D in Germany, which would undermine efforts for conversion and disarmament in the longer run.

EXAMPLES OF THE CONVERSION OF R&D

As mentioned above, a planned and integrated conversion strategy for Germany does not exist. This is especially true for R&D conversion, which is not seen by the Federal government as a major problem. In general, there is little interest in the conversion of R&D and adaptation to the changed situation is mostly left to the individual institutions. However, tentative approaches to deal with the problem of conversion do exist. The Bonn International Center of Conversion (BICC) was established in April 1994 by the state government of North-Rhine Westphalia with the assistance of the UN. One of the areas of activity will be focused on 'the reorientation of military research and development (R&D) facilities and making this R&D knowledge available for non-military purposes,' and there are the efforts by the German parliament (Subcommittee on Disarmament and Arms Control) in regard to preventive/qualitative arms control, which in the long-run could also include considerations of conversion of military-relevant R&D. These activities started in late 1992 and have led to a study project of the 'Büro für Technikfolgenabschätzung beim Deutschen Bundestag' (TAB, the Office for Technology Assessment at the German Parliament) on arms control criteria that is still underway.

Unavoidably, military-related R&D must adapt to the post-Cold war environment. The example below illustrates how the structure and the objectives of military-oriented institutions are influenced by the end of the Cold War and how efforts are undertaken to initiate conversion.

Fraunhofer-Institut für Angewandte Materialforschung (IFAM)

Bremen, the smallest German Federal State, has in the past been very dependent on arms production. In 1991, its arms industry consisted of ten major companies (with more than 100 employees) and several smaller ones. In the large companies, on average, some 30 per cent of the workforce was dependent upon defense for their employment. Those jobs were spread across a number of different industries. For the ten major companies, 40 per cent of defense employees were employed in electronics, 31 per cent in military shipbuilding and repair, around 25 per cent in military aircraft and vehicle construction/repair and 4 per cent miscellaneous.[43] Bremen also has a significant number of military bases and establishments. Perhaps 10,000 of the approximately 25,000 defense related jobs in the state are at risk in the next few years.

To support the conversion process, a foundation for conversion ('Stiftung Rüstungskonversion') was set up, and a representative for conversion was appointed. In the following years, it is intended to save jobs by converting various branches from the production of military to civilian goods. Among others, this affects an institute for materials research, the Fraunhofer-Institut für Angewandte Materialforschung (IFAM). IFAM has been based in Bremen for more than 20 years. Until recently the institute has been almost totally dependent on BMVg for funding, but now it is attempting to become less reliant upon defense R&D by seeking civilian projects. The research, done by 130 scientists, technicians and other employees, focuses on the improvement of advanced materials, composites and the engineering construction of defense equipment. The main expertise has been in technology for the penetration of and protection against conventional munitions, thus combining the development of new weapons with research on countermeasures against them. Special adhesive techniques, powder metallurgy and casting practices for composites and advanced materials are fundamental methods that were introduced and improved in Bremen. Most of these procedures are well developed now, but are not efficient or applicable economically.

In 1990 it was decided by the federal government and the state of Bremen that the IFAM should no longer get its basic funding from the BMVg but from the BMFT (30 per cent of the budget of the institute), phasing out after five years.[44] Now the BMVg only funds specific projects whose results could be made available for the civilian market. IFAM is officially seen as a pilot project for the conversion of military research institutes in Germany. The state of Bremen is supporting the transition phase (1992-1996) with 11.3 million DM (about 10-15 per cent of the institute's budget). The management of the institute is optimistic that it can succeed in applying part of its know-how to civilian products. For example, metallic foams for light-weight construction and composite fiber materials for the optimization of rotor blades for wind energy are being developed. Nevertheless the research on light-weight construction, high-temperature ceramics etc. can be regarded as highly civil-military ambivalent.[45]

One of the main problems for conversion of such an institution is that weapons researchers, who for a long time had privileged positions, now have to compete in existing civilian markets. Under the conditions of defense research (e.g., secrecy, command structure, no presence on the open market), the name and expertise of the institute were not well known in the industry. Today, there is a strong competition in the materials sector, and IFAM employees are not used to taking into consideration the diversity and interests of their customers; they have had no experience in how to acquire orders and civilian contracts.

An agreement for cooperation has been reached with the University of Bremen, despite some problems with the university law due to the civilian orientation of the university. The deal provides that IFAM directors can become professors at the university and that studies of students and diploma theses can be performed at IFAM. The institute is obliged to report annually on the status of the conversion process and to restrict its university-based work to exclusively civilian purposes. The fulfillment of these 'essentials' is a precondition for further cooperation after five years.

It is not obvious that the aim of irreversible conversion of IFAM is achievable. The BMVg aims to establish a precedent that could be attractive for other defense-related research institutions. Given the basic assumption that applied R&D will be dual-use capable, it is not yet clear whether this first R&D conversion example will lead to a totally independent civilian research institute or an institution which is partially doing military-oriented research.[46] Nevertheless this case deserves support by the public if the institute and its personnel are to succeed in converting its military character to civilian purposes.

CONCLUSIONS

Germany's dual-use strategy for R&D could lead to an emerging use of civilian technologies for military purposes, thus undermining efforts at irreversible conversion of R&D. Dual-use concepts are particularly problematic because they could counteract the conversion process by pursuing military interests while simultaneously seeking economic competitiveness in the civilian market.

Transparency in military-related R&D is an important prerequisite for a comprehensive program on efficient conversion. The lack of information and the secrecy about detailed budget figures and the purpose and institutions of military-related research are serious obstacles to pursuing R&D conversion. To overcome the blurring of the lines between civilian and military paths of the scientific development, new tools of science and technology assessment have to be introduced and used by decision-makers in the national and international arms control communities.[47] An 'ambivalence analysis' could reveal nodal points, which permit the analyst to distinguish between different paths of scientific evolution, and could open alternatives for decision-making. External decision points also need to be identified to weaken the influence of the military on civilian R&D. Additionally, preventive arms control measures and restrictions on military-related R&D in the international sphere are necessary to overcome the shortcomings of an isolated and unreflective promotion of R&D, thus preventing new technological arms races, new arms exports, and any ongoing use of science and technology for future wars.

NOTES

1. In 1990 around 1.35 million soldiers were deployed on German territory, decreasing by half to around 570,00 at the end of 1994. K. Francke, 'Die Senkung der Verteidigungsausgaben und ihre sozialen und wirtschaftlichen Folgen,' *NATO Brief* (August 1994), p. 14.
2. See H. Wulf, 'Konversion – wirtschaftliche und soziale Folgen,' *Spektrum der Wissenschaft* (September 1994), pp. 96ff.
3. See K. Francke, 'Die Senkung,' pp. 13-16. See also H.-J. Gieβmann, ed., *Konversion im vereinten Deutschland. Ein Land – zwei Perspektiven* (Baden-Baden: Nomos-Verlag, 1992).
4. See chapter 15 this volume.

5. More about conversion in *Abrüstung und Konversion. Politische Voraussetzungen und wirtschaftliche Folgen in der Bundesrepublik*, ed. L. Köllner and B. Huck (Frankfurt a. M./New York, 1990). For a comprehesive literature review see U. Albrecht, *Rüstungskonversions-forschung; eine Literaturstudie mit Forschungsempfehlungen* (Baden-Baden: Nomos-Verlag, 1979).
6. Whereas the last Russian troops left in 1994, in the NATO framework NATO troops are still deployed on German territory. In 1994 these were: United States (100,000), France (15,000), Netherlands (3,000), United Kingdom (38,200). See *Military Balance 1993/94,* (London: International Institute for Strategic Studies, 1994).
7. *Verteidigungspolitische Richtlinien* (Bonn: BMVg, 26 November 1992).
8. Initially this build-up was supported by technology imports, mainly from the United States. See H. Wulf, 'Western Europe: Facing Over-capacities,' in *Arms Industry Limited*, ed. H. Wulf (Oxford: Oxford University Press, 1993), p. 145.
9. A. Fischer, 'Zur Finanzlage der Verteidigungsausgaben,' *Wehrtechnik*, no. 9 (1993), pp. 5-6.
10. Note that some defense-related expenditures are hidden within other budget items.
11. K. Francke, 'Die Senkung,' p. 14.
12. K. Francke, 'Die Senkung,' p. 14.
13. H. Wulf, 'Western Europe: Facing Over-capacities,' pp. 96-101.
14. See 'Towards an European Armament Agency,' *Military Technology* (June 1994), pp. 10-20.
15. 'Co-operation is the only way of coping with the steady shrinkage in national defence budgets, notably in the fields of space, strategic transport, logistics and telecommunication.' From the address given by the WEU Secretary-General, W. van Eekelen, to the Royal Institute for International Relations, Brussels, 2 January 1994, in *Military Technology* (June 1994), p. 20.
16. W. Burr, 'Der Rüstungsbereich im sich wandelnden Umfeld,' *Wehrtechnik* no. 4 (1992), p. 5.
17. Erhard Heckmann, 'Luft- und Raumfahrt in der Krise,' *Wehrtechnik* no. 6 (1993), pp. 17-25.
18. Heckmann, 'Luft- und Raumfahrt.'
19. H. Wulf, 'Western Europe: Facing Over-capacities,' pp. 96-101. Continuing over-capacity has forced DASA to shrink in size. See Pierre Sparaco, 'Billions in Losses Force DASA Job Cuts,' *Aviation Week & Space Technology* 143 (30 October 1995), pp. 26–27.
20. I. Anthony et al., *Arms Production and Arms Trade, World Armaments and Disarmament, SIPRI Yearbook 1993* (Oxford: Oxford University Press, 1993), p. 444.
21. E.J. Laurance, S.T. Wezeman, and H. Wulf, *Arms Watch. SIPRI Report on the First Year of the UN Register of Conventional Arms*, SIPRI Research Report No. 6 (Stockholm: SIPRI, 1993). See also M. Chalmers; E.

Laurance, and H. Wulf, eds., *Developing the Register of Conventional Arms*, Bradford Arms Register Studies No. 4 (Department of Peace Studies, University of Bradford, 1994).

22. Defense Minister V. Rühe in an interview with *Süddeutsche Zeitung*, 8 May 1992.
23. Bundesministerium der Verteidigung (BMVg), *Weißbuch 1994* (Bonn: BMVg, 1994), p. 103.
24. Speech of V. Rühe at the Bundeswehr-Führungsakademie Hamburg, 16 December 1992, p. 15.
25. Bundesministerium für Forschung und Technologie (BMFT), *Bundesbericht Forschung 1993* (Bonn: BMFT, 1993). In the years 1987-1989, Germany's R&D share of the GDP was ahead of the United States.
26. BMFT, *Bundesbericht Forschung 1993.*
27. A critical analysis of the 1993 Federal R&D budget is given in R. Rilling and J. Stadlinger, 'Die Sprache der Ökonomie, Ein erst Bewertung des 'Bundesbereicht Forschung 1993',' *Dossier, Forum Wissenschaft*, no. 3 (1993).
28. The FhG is discussed as a potential model for U.S. R&D conversion in U.S. Congress, Office of Technology Assessment, *Defense Conversion: Redirecting R&D* (Washington, DC: U.S. Government Printing Office, 1993), Appendix A.
29. According to the then Minister of Science and Research, Gerhard Stoltenberg as reported in *Press Service of the BMWF*, no. 6 (1967).
30. See E. Bulmahn, 'Volkvertreter ohne Information. Transparenz - ein Fremdwort in der militärischen Forschung und Entwicklung,' in *Die Januskö pfigkeit von Forschung und Technik - Zum Problem der zivil-militärischen Ambivalenz*, ed. W. Liebert, R. Rilling, and J. Scheffran (Marburg: Bund demokratischer Wissenschaftler, June 1994).
31. Deutscher Bundestag, 'Die Entwicklung der Ausgaben für militärische Forschung und Entwicklung sowie für Friedens- und Konfliktforschung' (Response of the Federal government to a parliamentarian request by the SPD), Bundestagsdrucksache 11/7373 (Bonn, 12 June 1990), p. 20. Other items in the budget of the BMVg increase this by another 70 million DM. Additionally, 146 million DM from the budget of the BMFT should be added, since it has been declared military R&D spending under NATO criteria. See Bulmahn, 'Volkvertreter ohne Information.'
32. Bulmahn, 'Volkvertreter ohne Information.'
33. R. Schreiber, 'Wehrtechnische Forschung und Technologie im Europäischen Rahmen (EUCLID),' *Wehrtechnik* no. 12 (1990), pp. 76-92; P. Kraus, 'Die Luftwaffe und das FuT-Konzept der Bundeswehr,' *Soldat und Technik* no. 6 (1993), pp. 340-46.
34. BMVg, *Weißbuch 1994* (Bonn: BMVg, 1994), p.106.
35. BMFT, *Bundesbericht Forschung 1993*, p. 265.
36. For details see Schreiber, *Wehrtechnische Forschung*; and I. Davis, *Military R&D in Europe. Collaboration without Control?* Current Decision Report, No. 11 (Oxford: Oxford Research Group, October 1992).

37. J. Heyden, 'Forschung und Technologie im Rüstungsbereich,' *Wehrtechnik* no. 4 (1992), pp. 13-17.
38. See D. Fondran, 'Bündelung verteidigungspolitischer Forschungs-und Technologieresourcen,' *Wehrtechnik* no. 11 (1989), pp. 40-41; A. Seiler, *The Civilian-Military Ambivalence of High-Technologies and its Significance for Transnational Cooperation in Western Europe*, IANUS Working Paper 11 (Darmstadt: IANUS, 1992). EUREKA: European Research Coordination Agency; ESPRIT: European Strategic Programme for Research in Information Technology; EUCLID: European Co-operation for the Long Term in Defence.
39. *Wehrtechnik* no. 7 (1986), p. 61.
40. BMFT, BMWi, *Zukunftskonzept Informationstechnik* (Bonn: BMFT, BMWi, 1989).
41. Deutscher Bundestag, 'Die Entwicklung der Ausgaben,' p. 20.
42. Heyden, 'Forschung und Technologie.'
43. W. Elsner and G. Voss, *Bericht zu den Abrüstungsfolgen für das Land Bremen und zu den Handlungsmöglichkeiten* (Bremen: Ausschuß für Wirtschaftsförderung, 1991), p. xi.
44. H.D. Kunze, 'Konversion eines Instituts der Fraunhofer-Gesellschaft,' *Spektrum der Wissenschaft* (September 1994), pp. 104-07.
45. Cf. Statement of the IFAM director H.D. Kunze before the Subcommittee on Arms Control and Disarmament of the German Parliament. Deutscher Bundestag, 12. Wahlperiode, Unterausschuß für Abrüstung und Rüstungskontrolle, Protokoll no. 22, 9 November 1992.
46. Kunze, Statement before the Subcommittee on Arms Control.
47. See W. Liebert and G. Neuneck, 'Civil-Military Ambivalence of Science and the Problem of Qualitative Arms Control - An Example of Laser Isotope Separation,' in *Controlling the Spread of Military Technology. Lessons from the Past and Challenges for the 1990s*, ed. H.G. Brauch, H.J. van der Graaf, J. Grin, and W. Smit (Amsterdam: VU University Press, 1992), pp. 43-57.

10

Conversion of R&D: The UK Case

Tom Milne

INTRODUCTION

The United Kingdom displays a range of military capabilities matched only by the United States, Russia and France. The UK government spends roughly £2000 million on defense R&D, which is about 40 per cent of total government funding of R&D. It employs more than 15,000 people in its national defense laboratories, constituting around 40 per cent of the total number of people engaged in R&D for the government.[1] The UK's arms industry produces almost ten per cent of manufacturing gross domestic product, generates huge earnings from exports and employs about 400,000 people.[2] As these figures show, the UK's former imperial role has bequeathed strong armed forces and a big defense-industrial base, both of which have been slow to decline.

Defense cutbacks are being implemented in response to the end of the Cold War. The defense budget fell by over 10 per cent in real terms between 1992/3 and 1996/7 and defense R&D expenditure may fall to two-thirds of its Cold War level by the turn of the century.[3] In absolute terms though, spending will remain high. The United Kingdom deployed large forces to the Falklands in 1982 and to the Gulf in 1990/1, on each occasion with resounding military success, so the armed services can defend their core funding from a strong position. Moreover, although the threat that the Soviet Union was perceived to pose has all but vanished and scenarios in which the United Kingdom would today undertake solo military action are both few and improbable, it may well wish to take part in multinational and UN military actions in the future. In line with this the 1994 Defence White Paper states: 'The need to retain battle-winning technological advantage is as pressing now as it has ever been in the context of changing, unpredictable risks.'[4]

The UK's nuclear forces and supporting infrastructure are being scaled back. The UK is in the process of withdrawing its free-fall nuclear bombs, and has cancelled development of a tactical nuclear weapon system. The UK plans to deploy a submarine-based nuclear force for the foreseeable future, however.

Although there is no defense R&D conversion going on in the United Kingdom on a significant scale, the magnitude of UK defense work makes conversion an important subject to study. This chapter addresses the potential for the conversion of UK R&D.

GOVERNMENT FUNDING OF DEFENSE R&D

Table 10.1 shows the level of government expenditure on defense R&D over the last decade. Government establishments (notably the Atomic Weapons Establishment, the Defence Research Agency and the Chemical and Biological Defence Establishment) carry out the greater part of the research work, while industry does the majority of the defense development work. Universities do not have a major role. As can be seen, military R&D funding has fallen substantially in real terms since 1984. This is mostly as a result of reforms of the procurement practices of the Ministry of Defence (MoD) and a government decision, taken in 1987, to reduce the proportion of its R&D spending that goes to defense.

THE ATOMIC WEAPONS ESTABLISHMENT

The Mission Statement of the Atomic Weapon Establishment (AWE) is:

> To provide, maintain and certify the UK's nuclear deterrent, and to decommission redundant weapons, all in a safe and cost-effective manner.

The AWE's principal sites are at Aldermaston and Burghfield, both in Berkshire. The workforce at AWE is about 5000, of which more than one-third are professional scientists and engineers. By UK standards, the AWE is a large employer of scientists and engineers from a broad range of disciplines. The AWE's budget varies depending on the program of work, but is around £300 million per year.[5]

The AWE is owned by the Ministry of Defence but operated by the private sector consortium Hunting Engineering, Brown & Root and AEA Technology (Hunting-BRAE). Hunting-BRAE has a seven-year contract running from April 1993. The workforce will be employees of AWE plc, a subsidiary of Hunting-BRAE, for the duration of the contract.

Table **10.1**
MoD Expenditure on R&D, 1984/5–1996/7

	Net MoD expenditure on R&D: £ million: outturn prices	Net MoD expenditure: on R&D: £ million: 1992/3 prices	Percentage of total government expenditure on R&D
1984/5	1877.7	2925.9	47.4
1985/6	2018.2	2979.6	47.5
1986/7	1941.5	2784.8	45.5
1987/8	1910.7	2602.0	44.3
1988/9	1879.8	2398.8	42.8
1989/90	2028.6	2420.4	43.7
1990/1	2062.9	2277.8	42.7
1991/2	2267.0	2354.0	44.9
1992/3	2220.9	2220.9	42.7
1993/4[a]	2461.1	2383.6	45.2
1994/5[b]	2314.6	2155.5	43.5
1995/6[c]	2283.6	2049.8	42.8
1996/7[c]	2085.6	1826.4	40.6

[a] Estimated outturn.
[b] Provision
[c] Planned expenditure
Source: *Forward Look of Government Funded Science, Engineering and Technology 1994: Statistical Supplement* (London: HMSO, 1994), pp. 19, 21, 23. Figures are exclusive of VAT but include superannuation. Prices in real terms: 1992/3 base year. Government GDP deflator, *Forward Look*, p. 12.

The future of the British nuclear deterrent lies with the Trident submarine, due to remain in service until at least 2020. By mid-1998 the Trident missile will be the UK's only nuclear weapon. Until 1993 there were plans to develop a warhead for a tactical air-to-surface missile but this project has been cancelled. Thus, the end of the Cold War has removed the AWE's central task. It would now seem that the Aldermaston laboratory is most unlikely to be asked to design a new warhead for many years if ever. Its only nuclear weapons work will be decommissioning warheads retired from service and maintaining the safety and reliability of a small arsenal of modern weapons.

The comprehensive test ban treaty (CTBT) adds to the AWE's problems. The MoD has warned that testing has been an integral part of the AWE's research program and that there is a danger that nuclear expertise will attenuate in its absence. At least one independent commentator questions whether the AWE will be able to continue to recruit first-rate scientists after a test ban.[6] Various theoretical and experimental replacements for nuclear testing are being developed, and the House of Commons Defence Committee has urged that this work be treated as a priority:

> [I]t is vital that MoD assigns to the AWE the resources and profile necessary to ensure that nuclear expertise, and thus nuclear safety, can . . . be maintained without recourse to nuclear tests.[7]

The MoD is consulting with French officials about cooperation on warhead stewardship, including the possibility of joint work at the AWE and joint use of the AWE's facilities. This will be in addition to the longstanding cooperation that the UK enjoys with the United States. But although developing replacements for testing is an area in which the AWE's workload will expand, this may be only a palliative to the AWE's diminishing overall operation.

France's nuclear weapons laboratories are the closest foreign equivalent to the AWE. They are to reduce their 5700 staff by one-fifth over the next decade in a major reorganization. Two research centers will close down, including Limeil-Valenton, the main warhead design center, comparable to Los Alamos or Lawrence Livermore in the United States. Many of the scientists will transfer to the laser laboratory being constructed near Bordeaux.

There are no plans for a comparable reorganization at Aldermaston, although there has been considerable rationalization across the AWE, set to continue into the next decade. From a staffing level of 6400 in 1994, the current Chief Executive of Hunting-BRAE expects a workforce of around 4000 by the turn of the century.[8] Looking further ahead, a continuing reduction in funding for the AWE is a possibility. Large cuts have been made to the UK's conventional forces in recent years, and there will be strong resistance to further reductions. This might make the nuclear weapons budget a tempting target for future savings.

In 1993 the then Chief Executive of Hunting-BRAE, Brian Richards, said that the AWE had a fairly stable and well-defined program for the following 6–8 years. Looking beyond this period, Richards was doubtful about the prospects for substantial diversification, according it a lower priority than the efficient and safe delivery of the AWE's nuclear weapons remit. He said, 'We're going to be a nuclear establishment, principally, and if work for non-nuclear, non-MoD customers was to get to 10 per cent of our turnover, we would be doing pretty well.'[9]

At that time the priority at the AWE was to remedy problems and delays that had arisen in the Trident production process. As the Trident programme draws to a close, greater emphasis can be put on the laboratory's long-term future. Indeed, while it is obvious that Aldermaston will continue to exist in some form for as long as the UK has nuclear weapons, the Defence Committee believes that the AWE might *have* to diversify:

> It would seem inevitable, given the changes to the research programme and the reduction in production demands over the next decade, that if the AWE is to be maintained in a viable condition to preserve UK independent nuclear capability it must either be resourced – once Trident's needs have been met – beyond its immediate requirements, or other work than it is presently conducting must be found for it . . . without some form of future diversification, and maintenance of a demanding research programme, it will become difficult to justify the scale of public funding for the AWE such as to maintain the capability to sustain the maximum standards of safety and efficiency . . .[10]

One promising area that has been neglected by the AWE is the technical aspect to devising, implementing and monitoring nuclear arms control agreements and disarmament measures. Arms control work at the AWE has already included contributions to the definition of IAEA safeguards, seismological research for monitoring testing, radiochemical and microanalytical analysis for detecting proliferation, mathematical modelling of nuclear-weapon designs that might be used by proliferators, and the development of controlled access procedures for challenge inspections under the CWC. Although the AWE has the competence to contribute a lot to this field, it currently accords a low priority to such work, deploying only 1 in 500 of its staff on it.

The neglect of verification and arms control work in Britain contrasts strongly with the situation in the United States, where the nuclear weapons laboratories are spending hundreds of millions of dollars annually in this area. To correct this imbalance the AWE might seek to become a center of expertise on verification. It might even serve as a European center (there is little nuclear verification and arms control work taking place elsewhere in Europe) and be funded by the European Union as such. Picturing the AWE as a profitable part-converted defense establishment might appeal to a British government.

Verification work might also appeal to the AWE's staff. There was great enthusiasm in the workforce for diversification to medical technology work and other civilian areas that took place in the 1960s. Many nuclear weapons scientists have been motivated in their work by feelings of national loyalty. There could be similar enthusiasm for developing verification technology if this became a vital national interest.

A diversification process could also help to reform working practices and increase openness at Aldermaston. This is important for a number of reasons. Openness in nuclear work is essential for progress in disarmament, and would raise the quality of debate on nuclear defense. Greater openness would also allow independent peer review of scientific standards at Aldermaston. Many of the senior scientists at the US laboratories spend part of their time on unclassified research and participate in international conferences. The contributions of these scientists to 'open' science can justify confidence that their handling of nuclear weapons is of a high standard. The same assessment cannot be made in the UK - outsiders have no choice but to accept government assurances at face value. Lastly, openness will make it easier for Hunting-BRAE to recruit first-rate scientists, as it is attractive to be part of a broad academic environment and to publish in the open literature.

Thus, increasing work on verification and arms control at Aldermaston can be expected to appeal to several shades of opinion on British nuclear weapons - Aldermaston has resources and the expertise to work in this area and needs to diversify. Just one note of caution is appropriate. Expanding verification work must not be seen as functioning to sustain Aldermaston as a nuclear weapons laboratory through a period in which there is no nuclear weapons development. It is part of a disarmament process. But it surely would be viewed in this way if undertaken with due seriousness of purpose. In this light, it would be a significant conversion of military R&D.

THE DEFENSE RESEARCH AGENCY

The key purpose of the UK's conventional defense research program is to provide the Ministry of Defence with the expertise it needs to procure and maintain equipment that has a war-winning capability. The 1993 Science and Technology White Paper stated that 'as the Gulf conflict illustrated, technology can provide the decisive edge in military operations; a substantial investment in defense research and development is therefore vital for our armed forces.'[11]

The four major research establishments have traditionally been known as the Royal Aerospace Establishment (based at Farnborough in Hampshire), the Royal Armaments Research and Development Establishment (based at Fort Halstead in Kent), the Royal Signals and Radar Establishment (based at Malvern in Worcestershire) and the Admiralty Research Establishment (based at Portsdown, also in Hampshire). These establishments, which themselves incorporate a number of smaller MoD-owned sites, were amalgamated in 1991 to form the Defence Research Agency (DRA) and renamed DRA Farnborough, DRA Malvern, etc. The DRA's function is to provide independent scientific and technical services, primarily to the Ministry of Defence, but also to other customers in the public and private sectors.

The DRA was formed at a time when defense spending was falling substantially, and its viability depended on reducing costs while maintaining the level of scientific work. A Chief Executive, recruited from industry, was appointed and a process of rationalization was instituted that led to the closure of a number of sites and the loss of 2000 staff from a workforce of around 11,500. Even after these reforms, the DRA remains one of the UK's biggest employers of scientists and engineers. The DRA owns and manages a number of experimental facilities also used by industry and academia. These are now run as commercial concerns wherever this is possible. Where a facility is not commercially viable, but is a valuable national resource, the DRA keeps it open and receives compensation from the government.

Since April 1, 1993 the DRA has been managed as a Trading Fund - UK government jargon for a public sector body that has a contractor/customer relationship with government and is run to the greatest extent possible on commercial lines. No funds are voted to the Agency directly from government; instead it must tender for research work commissioned by individual budgetholders within the Ministry of Defence, other government departments or private sector clients. It is the MoD's policy to put contracts out to open competition in the market whenever feasible, and it anticipates that this will lead to a reduction in the proportion of research performed intramurally. (In the past, much of this work would have passed to government laboratories automatically.) The DRA is trying to compensate for reductions in funding from the MoD - which the DRA anticipates will continue - by increasing its work for other clients.

The management of the DRA as a Trading Fund achieved most of the government's objectives (by reducing costs, etc.), and in 1995 it was decided to amalgamate the DRA with the other major non-nuclear defense research agencies to form the Defence Evaluation and Research Agency (DERA). The new organization is constructed around the framework, procedures and financial systems of the Defence Research Agency. It is the largest science and technology organization in Europe with a turnover of over £1 billion.[12]

The DRA's remit covers land, air, and sea equipment. A review of potential requirements for post-Cold War military operations has found that there should be a shift away from research on major defense platforms towards 'smart' weaponry and C^3I equipment to be incorporated into more mobile and flexible armed forces. This will provide more work for some DRA sites than for others. There has also been increased consideration given to the linkages between military and civilian R&D, which may be significant for the conversion of R&D. In the past, the Ministry of Defence has had no remit to ensure that defense research benefitted the civilian economy. This has changed, and the new MoD philosophy is that its research program *does* have a role to play in promoting national prosperity. (In part this refers to revenue accrued from arms exports, but it also refers to technical advances that benefit civilian industry.) Increasing military-civilian cooperation is also motivated by the reverse side of the coin, that is, the benefits and savings that will accrue to the MoD from access to work funded in the civilian sector. However, although mechanisms have been put in place to increase technology transfer - indeed, the DRA's 1993 Framework Document gives it the explicit remit to cooperate with industry, and the DERA Framework Document commits the organization to seek to maximize wealth creation from its research activities - this remains a secondary objective of policy to be accomplished in ways that support the primary defense program.

The DRA promotes technology transfer in a number of ways. First, it places a substantial fraction of its research budget in research contracts in industry and academia. Second, it licenses patents to industry. Third, a scheme called Pathfinder, initiated in 1992, aims to inform firms as early as possible about the DRA's research program. The objective is to give companies the maximum opportunity to contribute to the DRA's work from their existing military or civilian research programs. Fourth, the Strategic Alignment Initiative, also begun in 1992, arranges for the mutual exchange of information on research programs between the DRA and about 30 big industrial firms through contacts between high-ranking executives. The aim is to create a strategic coherence in the national R&D base through rationalization of research programs to remove duplication of work, sharing of information on (promising and unpromising) lines of research, and joint analysis of long-term technological trends. Further initiatives are planned.

Another development within the DRA that is assisting technology transfer is the establishment of Dual Use Technology Centres. By centralizing facilities and workers, the DRA is trying to create, at a national level, effective and financially viable bodies of researchers and equipment. Industry and academia are able to use the facilities on a contract basis and office space is allocated to them. In this way the government hopes to gain greater leverage from defense resources, facilitate technology transfer for defense and civilian purposes, and make savings from rationalizing its facilities. The first such center, opened in April 1994, was the Structural Materials Centre (SMC), which consolidated DRA's materials laboratories, which had been spread throughout the South of England, at its Farnborough site. The second was the Farnborough Supercomputing Centre, opened in 1995. Other centers are being established to develop software engineering, maritime technology and electronics.[13] The DRA has identified 13 other areas in which it believes that its defense research has civilian applications (see box). These developments, which are of course additions to a long-established network of official and informal contacts, should indeed reduce duplication of research work in the United Kingdom, increase technology transfer between the defense and civil sectors and in general give the MoD better value for its expenditure on research. However, although an increase in dual use work may put defense research onto a more efficient footing and allow a reduction in the defense research budget, it will not, by itself, reduce the amount of defense equipment being designed and procured, and therefore does not fulfil the foremost aim of conversion. Indeed, it may even have the

DUAL USE AREAS IDENTIFIED BY DRA

air and surface vehicle traffic management systems
aircraft, helicopters and aero-engines
automated manufacturing and advanced structures
avionics, vetronics and control systems
environmental monitoring
imagery and display systems
infrared sensors and thermal imagers
navigation
parallel computing
radar
satellites
secure, high integrity and safety critical systems
surveillance and security systems
telecommunications and mobile communications

Source: Defence Research Agency, *Third Report of the House of Lords Select Committee on Science and Technology* (London: HMSO, 1994)

opposite effect and serve rather as a means of preserving the defense base in the face of declining government support and procurement. The advantage of a dual technology base from the standpoint of conversion is that it may put the laboratories into a stronger position to move into civilian areas should their defense work diminish.

International collaboration in defense research is expected to increase, especially within Europe. In addition to the benefits that derive from technical commonality between allies' defense equipment, the government has found that collaborative research projects can increase the return on investment in research.[14] The House of Lords Select Committee on Science and Technology has urged the MoD to cooperate with European NATO member states by designating specialized centers of excellence. It considers that the DRA will benefit from such a reorganization on a European level, provided account is taken of past records of scientific work and the facilities available.[15]

Although the DRA has substantially increased its links with industry and academia, and plans to continue to do so, the scale of non-MoD business, for the time being at least, will be modest. The DRA aims to develop £100 million a year of non-MoD business by 1998/9, some 15 per cent of turnover. Moreover, there is a limit to the dual use work that a defense research establishment can undertake. The majority of military research and most development work does not have any applicability to civilian areas. A report from the Advisory Council on Science and Technology estimated that in 1986/7 only about 20 per cent of the research and development budget was spent on technologies with potential applications in both military and civilian areas.[16]

PORTON DOWN CHEMICAL AND BIOLOGICAL DEFENSE ESTABLISHMENT

Porton Down, located in Wiltshire, is part of the Protection and Life Sciences Division of DERA. Porton has a staff of around 600, including 350 scientists. Its income for 1996/97 was £40 million. Porton describes its central purpose as 'protection against chemical and biological warfare through scientific understanding.' This involves work to assess potential chemical and biological weapon (CBW) threats; detect, identify, and monitor CBWs in the field; develop protective equipment; manage contaminated areas; develop medical countermeasures and treat CBW injuries. It also involves work on verification and arms control, relating, in particular, to the chemical and biological weapons conventions.

Porton has the equivalent of around 25 staff working full-time on chemical and biological arms control. Some of the staff spend only part of their time on arms control work, so the total number of Porton scientists devoting a proportion of their time to this area is higher. The annual income of the arms control program is around £2 million. Although the arms control work at Porton comprises only a twentieth of its overall operation, this work is now described in some detail because, in some respects, it provides a model for the sort of arms control program that it might be desirable to establish at Aldermaston and in other government defense research establishments.

Porton Down advises the Technical Secretariat of the Organization for the Prohibition of Chemical Weapons (OPCW), which is preparing to implement the verification system provided for in the Chemical Weapons Convention (CWC). It is a designated laboratory with the OPCW, meaning that it can carry out support work for the CWC on an official basis. In addition, Porton houses the UK's Single Small Scale Facility, provided for in the CWC, which is permitted to manufacture under international monitoring chemical agents for protective purposes; it is preparing the declaration of past and present UK activities in the chemical weapons field that is required under the CWC; it is developing a variety of technologies for verifying compliance with the CWC; and it has begun commercial consultancy work, in partnership with Logica (UK) Ltd's Business Consulting Group, assisting industrial plants to prepare for CWC inspections.

Verifying the CWC will be a complex and ever-changing undertaking - chemical weapons are steadily becoming cheaper and easier to produce - and will require ongoing support from bodies such as Porton.

Porton is equally concerned with the Biological and Toxin Weapons Convention (BTWC). Although the BTWC entered into force in 1972 and has been signed by over 150 nations, there are still no adequate arrangements for monitoring compliance with its provisions. Porton sends representatives to ad hoc groups working on strategies to verify the BTWC. It is also developing verification techniques now, in advance of an official requirement, so that Britain is able to contribute to the development of a verification regime on the basis of a sound scientific understanding of the possibilities and limitations of the available technology. For example, technologies that produce false positive results, which might be satisfactory for detecting biological weapons on the battlefield, where a better-safe-than-sorry policy would be followed, are not suitable for politically sensitive international verification procedures.

Biological arms control is likely to be an important and expanding area of work. Biotechnology makes it possible to produce new biological weapons agents and to manufacture toxins that are not naturally available in militarily significant quantities. Moreover, biotechnology is a fast-advancing field of science. Techniques that are hard today will be easy in a few years' time.

Porton also provides technical advice on chemical and biological weapons issues to various departments in the UK government. It supports UK government delegations to international forums such as the Conference on Disarmament – it was the principal technical adviser to the UK delegation that negotiated the CWC; it makes technical contributions to controls on trade in chemical and biological agents; and it advises on scientific aspects of non-proliferation policy. Scientists from Porton have also taken part in UNSCOM missions in Iraq.

There would seem to be a continuing need and strong support for CBW arms control work. Indeed, the importance of this work has come to the fore since the Gulf War, and, more recently, the terrorist use of the Sarin nerve agent on the Tokyo underground. The 1994 Defence Costs Study, *Front Line First*, refers to Porton's 'world-wide reputation as a centre of excellence,' which will be safeguarded.[17]

Looked at from a different perspective, however, there is reason to speculate that funding for verification and arms control might fall in coming years. Before moving to the trading fund operation, Porton received an annual budget, and it was left to Porton's Director to divide the money between competing laboratory programs. In the new organization, budgetholders in the MoD have the money to spend at their discretion. A baseline figure for arms control funding, determined by the funding levels for the previous years, was established for the first two years of trading fund operation, but there have already been small reductions in funding caused by the general pressure on the defense budget. It would not be surprising were spending on arms control work to be reduced when future savings are sought from defense expenditure. Thus the level of political backing for international treaties such as the CWC and BWC will be crucial.

INDUSTRY

UK industry received £1296 million of government funds for defense R&D in 1992/3 and spent about £400 million of its own funds on it.[18] The Table below lists the contractors that received more than £50 million from the MoD in 1993/94. These firms do most of the defense R&D carried out in industry.

Table **10.2**
MoD Contractors, 1993/94

> £250 million	£100–250 million	£50–100 million
British Aerospace plc	Babcock International Group plc	British Petroleum Co plc
Building and Property Facilities Management Ltd	British Telecommunications plc	BTR plc
Devenport Management plc	IBM UK Holdings Ltd	Building Management Scotland Ltd
General Electric Co plc	ICL plc	ESSO UK plc
Hunting plc	John Mowlem and Co plc	Ferranti International plc
Rolls Royce plc	Siemens plc	GKN plc
VSEL plc	Thorn EMI plc	Leyland Trucks Ltd
other government departments	Trafalgar House plc	Northern Telecom Europe Ltd
	Vickers plc	Sema Group plc
	Westland Group plc	Serco Group plc
		Swan Hunter Ltd
		TBV Consult Ltd
		TI Group plc
		WS Atkins Ltd

Source: *Defence Statistics 1994* (London: HMSO, 1994), p. 17.

There are many obstacles facing a firm trying to convert from military to civilian work; these are discussed in earlier chapters of this book. In the past, firms that have attempted to convert their military work while operating in a market economy have usually failed. The majority of UK defense companies have reacted to the shrinking post-Cold War defense market either by consolidating their defense work through joint ventures and mergers, rationalizing their defense business, or divesting themselves of their defense divisions. Research carried out in 1991/2 for the Parliamentary Office of Science and Technology failed to find 'any example of successful complete conversion in recent years.'[19] This is supported, in the main, by a 1994 Report from Management Consultants Ernst and Young on the future of the UK defense industry, which found that insiders in the industry share a belief that what scope there is for conversion lies in the aerospace field alone, and that even in this area conversion will be restricted by the existing civil capabilities.[20] The government has sought to make it easier for firms to work in both the military and civilian sectors through reforms to MoD procurement practices and schemes such as Pathfinder but the primary objective of such policies is to preserve a viable defense-industrial base.

UNIVERSITIES

The MoD funds a relatively small portfolio of research in UK universities to a total value of £20-30 million a year, which is three to five per cent of the defense research budget. The research it supports is frequently sponsored jointly with one of the Research Councils. Although the MoD does not fund any basic research (by Frascati classifications) that by definition is not specifically relevant to defense, there is no intrinsic reason why a university scientist could not shift his or her research away from military applications. It would not require any major change in work practices or organization. The difficulty would be replacing the MoD support with funding from another source.

CONCLUSIONS

The projected one-third reduction in UK expenditure on military R&D through the 1990s will, in the main, be realized from i) new management practices, rationalization and other reforms within the government defense establishments; ii) increased use of technology developed in the civilian sector and increased international collaboration in defense research; and iii) the overall downsizing of the UK defense program.

There is little discussion about, and no plans for, conversion of the UK's defense research establishments.

In all probability, defense R&D will not be reduced rapidly beyond the levels now planned. A development that would not depend on the immediate conversion of the laboratories, but which would nevertheless be quite revolutionary, would be to establish operations analogous to the one at Porton Down on at least as large a scale (i.e., five per cent of staff) within the AWE and the DRA. This is within the range of currently feasible policies, would make a significant contribution to the causes of arms control and disarmament, and could turn out to be a precursor of a program of conversion in a broader sense (in the sense that Porton Down has been part of the process that may realize a chemical-weapon-free world).

NOTES

1. *Forward Look of Government-Funded Science, Engineering and Technology: Statistical Supplement* (London: HMSO, 1994).
2. *The UK Defence Industry - Securing Its Future*, Ernst & Young, 1994.
3. *Realising Our Potential: A Strategy for Science, Engineering and Technology*, Cm 2250 (London: HMSO, 1993), p. 39.
4. *Statement on the Defence Estimates 1994,* Cm 2550 (London: HMSO, 1994), p. 64.
5. Statement on the Defence Estimates 1994, *Sixth Report of the Defence Committee 1993–94* (London: HMSO, 1994), p. 45.
6. P. Lewis, 'What do you want with an old banger?' *New Scientist* (2 July 1994).
7. *Progress of the Trident Programme*, Second Report of the Defence Committee 1993–94 (London: HMSO, 1994), p. xvii.
8. *AWE News-Link*, No. 19, p. 3.
9. 'The Way Ahead for Aldermaston,' interview with Brian Richards, *Newbury Weekly News* (31 December 1993).
10. *Second Report of the Defence Committee 1993–94*, pp. xxiii-xxiv.
11. *Realising Our Potential*, p. 39.
12. Defence Evaluation and Research Agency, *Annual Report and Accounts 1995/96* (London: HMSO, 1996), p. 4.
13. Defence Evaluation and Research Agency Annual Reports and Accounts 1995/96, pp. 31–32.
14. *Statement on the Defence Estimates 1994*, p. 66.
15. *Defence Research Agency*, Third Report of the House of Lords Select Committee on Science and Technology (London: HMSO, 1994), p. 19.
16. Advisory Council on Science and Technology, *Defence R&D: A National Resource* (London: HMSO, 1989), pp. 15–16.
17. *Front Line First: The Defence Costs Study* (London: HMSO, 1994), p. 20.
18. *Forward Look of Government-Funded Science, Engineering and Technology,* pp. 28, 35.
19. *Conversion and Diversification of Defence Technology and Manufacturing,* Parliamentary Office of Science and Technology, 1992, p. 18.
20. *The UK Defence Industry - Securing Its Future*, p. 12.

PART III

Specific Opportunities and Challenges

11

Conversion of U.S. Nuclear Weapons Laboratories

Ray E. Kidder

INTRODUCTION

The Lawrence Livermore and Los Alamos National Laboratories are major U.S. government research and development organizations that have been primarily concerned with the development of nuclear explosives, the high-pressure and high-temperature states of matter they produce, and the broad areas of physical science necessary to their understanding. They are considered to rank among the top research establishments in the United States in comparison with the rest of the national technology base, and they command facilities that are unmatched at most companies and universities. Now that the need for further development of nuclear weapons has greatly diminished or indeed disappeared with the end of the Cold War, what should be done to best utilize the extensive, modern, but somewhat specialized capabilities of these Laboratories?

The main finding of this article is that the defense mission of the Laboratories will be continued, although changed in its focus, for the foreseeable future, making conversion issues less relevant than they would otherwise be.

THE LABORATORIES' DEFENSE MISSION

The primary responsibility of both the Lawrence Livermore National Laboratory (LLNL) and the Los Alamos National Laboratory (LANL) is their special role in defense, particularly in nuclear weapons technology.

All of the research, development, and testing of the nuclear explosive components of U.S. nuclear weapons is done by these two Laboratories. Responsibility for research, development and testing (RD&T) relating to the non-nuclear components of nuclear weapons lies with the Sandia National Laboratory.

Both nuclear design laboratories, Livermore and Los Alamos, are owned by the Department of Energy (DOE) and operated by the University of California (UC) according to the terms of a renewable five-year contract. The University has operated both Laboratories without interruption since they were founded many years ago: the Los Alamos Laboratory since 1943 and the Livermore Laboratory since 1952.

The five-year contract for each Laboratory was renewed in November 1992, and would normally be expected to remain in force until November of 1997. The preamble, nuclear defense, and non-nuclear defense portions of the mission statement of the Los Alamos National Laboratory, as set forth in Appendix K of the new DOE-UC contract, are as follows:

> The Los Alamos National Laboratory is dedicated to applying world-class science and technology to the nation's security and well-being. *The Laboratory will continue its special role in defense, particularly in nuclear weapons technology*, and will increasingly use its multidisciplinary capabilities to solve problems in the civilian sector (emphasis added).
>
> The Laboratory strives to serve DOE and other federal agencies, and to complement and collaborate with universities and industry, by solving problems that typically:
>
> - are large in scale of time, space, size or complexity;
> - require a strong science base;
> - require engineering, teamwork, and special facilities;
> - benefit from a multidisciplinary approach and continuity of effort; and
> - benefit from a public service orientation.
>
> Some of the key elements of the Laboratory's mission are as follows:

Nuclear Defense:

The Laboratory shall continue its special role of providing scientific and engineering support to national nuclear defense activities, including R&D relating to:

- nuclear weapons safety, security, and safeguards;
- safe and efficient build-down and maintenance of the nations nuclear weapons stockpile;
- international arms control, nonproliferation, detection and verification;
- inertial confinement fusion, including high energy laser development and target design and fabrication;
- reconfiguration of the DOE weapons production complex;
- nuclear materials processing, storage, recycling and disposal;
- preservation of core competencies related to nuclear weapons technology.

Non-Nuclear Defense:

The Laboratory shall continue R&D in selected areas of conventional defense technologies, including:

- advanced munitions and energetic materials;
- materials science and armor/anti-armor applications;
- analysis and simulation;
- nonnuclear strategic defense initiatives.

The defense portion of the mission statement of the Lawrence Livermore National Laboratory is similar and only slightly less extensive (see Appendix 1).

These mission statements make it clear that the defense missions of *both* nuclear design laboratories were in 1992 intended to be very broad in scope for at least the next five years, including 'activities associated with all phases of the nuclear weapon life cycle and related tasks.' As discussed below, however, there has been a virtually continuous process of review of laboratory missions and funding, with consequent instability in the laboratories. For example, Complex 21, a major program for creating a modernized weapons production complex that had been assigned to Lawrence Livermore National Laboratory, was quietly cancelled early in 1994 as it became accepted that any new weapons production was unlikely in the foreseeable future.[1]

RECENT LIMITATIONS ON NUCLEAR WEAPONS TESTING

On October 2, 1992, President Bush signed into law (P.L. 102–377) the FY 1993 Energy and Water Appropriations bill, which contained an amendment sponsored by Senators Hatfield, Mitchell, and Exon limiting nuclear weapons testing. This amendment imposed a moratorium on testing until July 1, 1993; permitted not more than 18 tests (6 per year) prior to October 1, 1996; and none thereafter unless another nation conducts a nuclear test. The amendment limited tests by the United Kingdom at the Nevada test site to a total of three (one per year) during this same period, and required that these tests be counted against the total of 18. If the United States chose to allow the United Kingdom to exercise this option, the United States would then be limited to a maximum of 15 tests, or not more than five tests in each of the fiscal years 1994, 1995, and 1996. The amendment further limited the nature of these five tests to four for improving weapon safety and one for improving or verifying weapon reliability.

On July 3, 1993, President Clinton extended the U.S. moratorium on nuclear weapons testing until October 1, 1994, when it was reviewed and further extended. He stated that he was confident that the U.S. arsenal was 'safe and reliable,' but also said that 'to assure that our nuclear deterrent remains unquestioned under a test ban, we will explore other means of maintaining our confidence in the safety, the reliability, and the performance of our weapons.' A Presidential Directive was issued, followed by Congressional approval of a budget line item for FY 1994 that established the 'Stockpile Stewardship Program' calling for

> the preservation of U.S. competency in nuclear weapons, including weapons design, system integration, manufacturing, security, use control, reliability assessment, and certification.[2]

That is, it calls for an across-the-board competency in nuclear weapons work as practiced by the weapons labs and the production complex. The Stockpile Stewardship Program thereafter provided the 'marching orders' for the Livermore and Los Alamos National Laboratories.

President Clinton warned, however, that any resumption of testing by another country would lead the United States to move quickly to conduct additional tests. Such a provisional 'no first test' policy does not necessarily imply the end of U.S. nuclear weapons testing, dependent as it is upon the willingness of other countries to forgo further (or first-time) nuclear tests. China, in particular, has indicated that, having conducted only 42 nuclear tests to date as compared to 942 tests conducted by the

United States and 969 by Russia, France and Britain combined, it may be unwilling to terminate its modest nuclear testing program unless the United States will commit itself additionally to a policy of 'no first use,' a policy the United States continues to oppose. These issues and others are being discussed by the Conference on Disarmament in Geneva in a series of negotiations on a Comprehensive Test Ban (CTB) Treaty.

In view of these and other uncertainties, both short-term and long-term, the U.S. plans not only to maintain its stockpile of nuclear weapons in good working order, but also to maintain a capability to resume nuclear testing on relatively short notice. What effect will these developments have on the defense mission of the nuclear design laboratories?

In a presentation to the UC President's Council on the National Laboratories on October 29, 1992 John H. Nuckolls, then Director of LLNL, projected a 40 per cent reduction in core nuclear weapons RD&T by FY 1998. However, he also projected that this reduction would be more than offset by increases in nuclear weapons-related activities such as non-proliferation, environmental restoration and nuclear waste management; planning for Complex-21, the DOE's modernized nuclear weapons production complex (subsequently cancelled); inertial confinement fusion (ICF); and the Nuclear Test Ban Readiness Program mandated by the Congress in FY 1989. In his State of the Laboratory address on September 2, 1993, Nuckolls forecast that 50 per cent of the LLNL 1998 budget would be devoted to defense, including: nuclear weapons stewardship, dismantlement, verification, non/counter-proliferation, and 21st Century non-nuclear weapons.

One can conclude that there will be a change in focus of the defense mission of the nuclear design Laboratories as a result of the test limitations imposed by the Hatfield Amendment and the current moratorium on nuclear testing, but currently-planned weapons-related activities not dependent on nuclear tests and mandated by the Stockpile Stewardship Program will help to insure that the overall funding of their defense mission remains relatively stable.

HOW TO DEFINE A 'NUCLEAR TEST'?

The Hatfield Amendment and the Nuclear Test Moratorium places limitations on nuclear explosive tests, but does not specify what constitutes a nuclear test. The clear intent of the Amendment and the Moratorium is to strongly limit the further development of nuclear weapons. On the other hand, it is generally agreed that the laboratory-scale thermonuclear micro-explosions that are the goal of ICF programs in a number of countries as well as the United States need not and should not be prohibited, because they are of only minor significance to nuclear weapons development and may possibly be important as a future energy source. How can these conflicting interests be resolved?

The largest nuclear explosive yield that need be permitted to accommodate the needs of ICF is a yield equivalent to that of a few hundred pounds of high explosive (HE). Fortunately, nuclear explosive yields associated with nuclear weapons or explosive sources of nuclear radiations are more than one hundred times larger, as indicated by the 'No Man's Land' gap between the Non-Fission-Driven Regime (i.e., ICF) and the Fission-Driven Regime shown in Figure 11.1.[3] The intent of limitations on nuclear testing would be satisfied without risking future interference with the ICF program if *a nuclear test were limited in yield to no more than a few hundred pounds of HE equivalent.*

Figure **11.1**
Comparison of Nonfission-driven and Fission-driven Testing Regimes

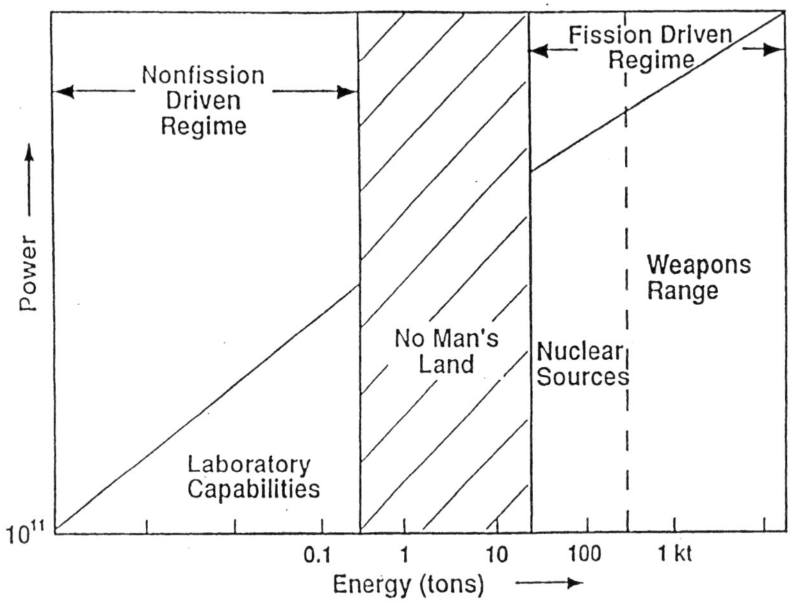

Further development of modern nuclear weapons would not be practical if so limited. A limit set at a few-hundred-pound yield, which is much smaller than that of some conventional aircraft-delivered bombs, is more than adequate to permit the one-point nuclear safety tests[4] that are required of all nuclear weapons in the U.S. stockpile. Such tests could be fully and safely contained in a properly designed above-ground facility should that be required.

Seismologists do not believe that yields as low as a few hundred pounds could be detected and identified as being of nuclear origin by teleseismic means. An explosive yield of at least ten tons (if fully coupled to the surrounding teleseismic medium), and at least a kiloton (if fully decoupled by being exploded in a sufficiently large underground cavity), would be the lowest yield for which teleseismic detection and identification would be effective. However, permitted low-yield explosions could be required to be conducted in containment facilities subject to inspection, and equipped with tamper-proof instrumentation to monitor the explosive yields produced.

As of this writing, a wide spectrum of yield limits have been proposed by the five declared nuclear weapons states: China (0 lbs), United States (4 lbs), United Kingdom (100 lbs), Russia (10 tons), France (100–200 tons, conditionally). It is not yet clear what yield limit will finally be agreed upon, but a limit as large as ten tons may be more strongly predicated on the difficulty of verification at lower yields and on facilitating one-point safety tests, than on limiting further weapons development (ICF yields would not be affected by specific limits on *fission* yields, however, whatever they might be).

POSSIBILITY OF CONSOLIDATING ALL NUCLEAR WEAPONS WORK AT ONE LABORATORY

In a letter of February 8, 1992 to the Secretary of Energy James D. Watkins, the Chairman of the House Committee on Science, Space, and Technology, George E. Brown, proposed, inter alia, that:

> A better solution – for the labs and for the nation – would seem to be found in a plan that gave the existing weapons labs [Livermore, Los Alamos, and Sandia] three distinct missions. Specifically, my proposal, for your consideration, would be the following:

1) *Consolidate all nuclear defense and nuclear non-proliferation work at Los Alamos.* Through a gradual transfer lasting perhaps three to five years, all nuclear design, ballistic missile defense, and other classified defense activities currently at Livermore would be terminated or transferred to Los Alamos, which would become the Los Alamos National Defense Laboratory. Nuclear weapons production activity, albeit at low production rates, could also be supported by existing Los Alamos facilities.

2) *Establish Lawrence Livermore as a civilian technology lab.* Building on strengths at the lab in material science, computational science, environmental remediation, and biotechnology, create a new, entirely civilian mission for Lawrence Livermore, with particular emphasis on building consortia with industry and academia in areas of critical technologies and environmental technologies. The lab could be renamed the Lawrence Livermore National Critical Technologies Laboratory.

3) *Concentrate verification activities at Sandia National Laboratory, in addition to making Sandia a 'center of excellence' for technology transfer.*

Congressman Brown's proposal to consolidate all nuclear defense and nuclear non-proliferation work at Los Alamos did not find favor either at the Livermore Laboratory or at the DOE. The argument was that such consolidation would eliminate the benefits of competition between the two labs and also the desirable checks and balances provided by the expert peer review that each lab provides to the other. Los Alamos indicated that it would also oppose the plan if the Laboratory were to be confined to defense work alone.

It is likely, however, that facilities at Livermore that are used for processing substantial quantities of hazardous radioactive materials will be closed. The tritium facility at Livermore has not been used for nuclear weapons work since early 1991 and its work has been transferred elsewhere. The DOE is currently considering closing the plutonium facility at Livermore and transferring its work to the larger facility that presently exists at Los Alamos. The location of the Livermore Laboratory in the midst of an encroaching suburban population, as well as the desire to consolidate hazardous nuclear weapons activities where possible, motivates the transfer of these Livermore activities to a less densely-populated area such as Los Alamos.

Brown's proposal received a new lease on life with the February 1, 1995 report of the Galvin task force, a blue-ribbon panel charged by the Secretary of Energy Hazel O'Leary with evaluating the future of the energy department's physics laboratories. The Galvin Panel (named for its head, Robert Galvin, chairman of Motorola Inc.), recommended consolidating nuclear materials development and production work at Los Alamos, but leaving Livermore missions in the area of nuclear nonproliferation, treaty verification and intelligence work. Other proposals were to establish a non-profit corporation to run the laboratories and to direct the laboratories to focus on basic research rather than the technology transfer activities that had burgeoned in the early 1990s. It is not clear what actions will follow from the Galvin panel's report, which is very controversial, but it appears that its findings will influence policy decisions about future missions and governance of the laboratories.[5]

BUDGETARY IMPLICATIONS

It now seems possible, but is not yet certain, that by October 1 of 1996 or perhaps earlier, a ban on nuclear testing of unlimited duration could be agreed to by the United States. The cessation of testing could be expected to yield annual savings of $300–350 million. (For FY 1995 the DOE requested $362 million for nuclear testing. Some of these funds would continue to be required to maintain the Nevada Test Site in readiness to resume testing without undue delay should a future need to do so arise.)

Taking Livermore as an example, the planned FY 1995 operating budget (in millions of dollars) was divided between nuclear defense, non-nuclear defense, and civilian missions as follows:

Table **11.1**
FY 1995 Operating Budget for LLNL

	M$	%
Nuclear Defense	333	38
Non-Nuclear Defense	91	11
Total Defense	424	49
Civilian Missions	443	51
Total FY 1995 Funding	867	100

A more-detailed breakdown of these costs is provided in Appendix 2 together with the corresponding costs for the preceding year. The total operating budget planned for FY 1995 was slightly larger (0.6 per cent) than the FY 1994 budget.

If the proposal to gradually terminate or transfer to Los Alamos all defense work at Livermore were adopted, annual savings as large as $350–450 million could ultimately result, the actual amount depending on how much of the Livermore defense work was terminated rather than transferred.

COOPERATIVE R&D AGREEMENTS

Whether terminated or transferred, the loss in defense funds at Livermore could be offset by an increase in funds for civilian missions. Cooperative Research And Development Agreements (CRADAs) between the weapons labs and private industry, in which costs are shared by the DOE and the industrial partner, are expected to be a significant component, eventually accounting for as much as 20 per cent of the total laboratory budget.

At the end of FY 1994, 128 CRADAs had been executed at Livermore, bringing a total of $466 million to the CRADA projects (to be spent over their two-to-five-year duration). Grouped by related industry, they were:[6]

Table **11.2**
Distribution of CRADAs at LLNL by Area of Technology, 1994

Manufacturing and Materials	30%
Computing and Communications	18%
Semiconductors, Microelectronics, Photonics	14%
Electronics Manufacturing	11%
Biotechnology, Health Care	8%
Modeling of Industrial Processes	7%
Environmental/Emission Reduction	7%
Energy	5%
	100%

CONCLUSIONS

The Cold War is indeed over, but the unsettled and potentially dangerous state of affairs that presently exists in the former Soviet Union and elsewhere suggests that any decrease in DOE defense funding at Livermore and Los Alamos is likely to be both measured and modest for the next several years. The DOE defense laboratories have pointed out that their RD&T budget represents only a modest fraction of the overall DOE defense activities budget. In FY 1994, for example, their RD&T budget was only 16 per cent of the DOE defense-related spending. The laboratories defend their budget on the grounds that it is cost-effective to fully utilize the leverage of their RD&T talents to reduce the far larger costs of current and future DOE defense activities mandated by the Stockpile Stewardship Program.

To carry out the Stockpile Stewardship Program, DOE plans[7] to continue to operate both of its weapons design labs (Livermore and Los Alamos) and its engineering lab (Sandia), and to construct several new facilities - including the Dual-Axis Radiographic Hydrotest facility for hydrodynamic tests and the National Ignition Facility (an ICF facility) at Livermore for research on the fusion portions of the weapons - to provide data on the reliability and safety of weapons as they age. In addition, DOE plans to keep the Nevada Test Site operational so that it can conduct hydronuclear experiments (hydrodynamic tests in which a very small nuclear explosion - equivalent to a few pounds of TNT - actually occurs). The three laboratories also plan to spend some $220 million annually for CRADAS and other technology transfer initiatives in which laboratory scientists work with industry to share technology with the private sector.

Significant reductions in defense RD&T expenditures at the weapons design laboratories will not materialize unless the currently planned scope of the DOE's Stockpile Stewardship Program is significantly reduced. An option described by the Congressional Budget Office[8] that would consolidate the two design laboratories, forgo all hydronuclear testing activities at the Nevada Test Site, trim funding for CRADAs by one-third, and gradually increase funding for other stewardship activities, would save an average of $184 million annually (only about 10 per cent of the present annual budget of the two design labs) for the five-year period from 1996 through 2000.

Although pressures are strong to cut federal expenditures in order to reduce the deficit and balance the budget, these pressures do not presently extend to national defense. It therefore seems unlikely that substantial reductions in the combined defense-related budgets of Livermore and Los Alamos will materialize for some time to come.

NOTES

1. Christopher Anderson, 'Livermore Faces Forces of Change,' *Science* 264 (15 April 1994), pp. 336–38.
2. *Congressional Record* (10 November 1993), p. H9304.
3. Figure presented by R.E. Kidder at a 1989 Congressional hearing. U.S. Congress, House of Representatives, Committee on the Armed Services, *The Nuclear Test Ban Readiness Program: Hearing Before the DOE Defense Nuclear Facilities Panel.* 101st Congress, 1st sess., 26 April 1989.
4. 'One-point safety' of a nuclear warhead requires that there be less than one chance in a million that detonation of the warhead's high explosive at any one point will result in a nuclear yield exceeding that equivalent to four pounds of high explosive.
5. See Andrew Lawler, 'Report to Stress Research over Close Ties to Industry,' *Science* 276 (27 January 1995), pp. 446–47.
6. CRADA data obtained from Lawrence Livermore National Laboratory 'FACT SHEET-FY94' (April 1995).
7. DOE plan as reported in U.S. Congress, Congressional Budget Office, *Reducing the Deficit: Spending and Revenue Options* (Washington, DC: U.S. Government Printing Office, February 1995), p. 24.
8. Congressional Budget Office, *Reducing the Deficit*, p. 25.

APPENDIX 11.1

MISSION STATEMENT
LAWRENCE LIVERMORE NATIONAL LABORATORY
(Preamble and defense portion)

The mission of the LLNL is to serve as a national resource in science and engineering, focused on national security, energy, the environment, biomedicine, economic competitiveness, and science and engineering education, with a special responsibility for nuclear weapons.

Mission programs include:

- Research, development, and testing (RD&T) activities associated with all phases of the nuclear weapon life cycle and related tasks.

- RD&T associated with non-proliferation, arms control, and treaty-verification technology.

- Inertial confinement fusion (ICF) for understanding weapons physics, providing weapons-effects research, and developing a practical commercial energy source for national security.

- Strategic defense research emphasizing kinetic- and directed-energy weapons, including the supporting space technology.

- Support of the Department of Defense (DOD).

- Support of the intelligence community.

APPENDIX 11.2

COSTS BY MAJOR PROGRAM
LAWRENCE LIVERMORE NATIONAL LABORATORY
($ in millions)

	FY 1994 Actual	FY1995 Estimate
Nuclear Weapons RD&T	195.8	173.4
Inertial Confinement Fusion	77.6	76.0
Non-Proliferation & Intelligence	66.4	59.1
Other DOE Defense	22.4	24.8
Total Nuclear Defense	362.2	333.3
Advanced Implementation Tech.	34.4	9.4
Other DOD	84.3	81.6
Total Non-Nuclear Defense	118.7	91.0
Advanced Isotope Separation	4.2	3.2
Env. Rest. & Waste Mgmt.	70.7	63.3
NER Supercomputer Center	38.7	44.0
Energy Research	29.4	37.8
Biomedical & Environmental	28.3	28.1
Magnetic Confinement Fusion	14.5	17.1
Basic Energy Science	8.1	9.0
Technology Commercialization	45.5	58.9
Other DOE Organizations	91.6	109.1
Other Non-DOE (w/o DOD)	49.9	72.2
Total Civilian Missions	380.9	442.7
Total Laboratory Operating	861.8	867.0
Construction/Equipment	107.1	110.2
Total Laboratory	968.9	977.2

12

Conversion Problems of Russian Nuclear Weapons Centers

Evgeny Avrorin

INTRODUCTION

The dramatic events of the last few years, such as the radical changes in the international situation, the end of the military confrontation between the two superpowers and the arms race this caused, the disintegration of the socialist community and then of the Soviet Union, and the destruction of the former political and economic system, have completely changed the situation in Russia's nuclear weapons centers, making it necessary to change the scientific subjects being studied as well as organizations, structures and personnel. The reduction in military orders and the shift in the Russian economy to meet the requirements of conversion have necessitated deep conversion of all military science and industry, including nuclear weapons centers. In terms of their scientific level, structure, and personnel and the position they held in the military-industrial complex of the former USSR, these centers were unique scientific and design institutions.

This chapter is devoted to the conversion of scientific research, design and technological work in the field of the problems of nuclear weapons - their special features, successes and blunders - from the perspective of an insider in the nuclear complex. In my opinion the many problems and difficulties related to conversion of the nuclear centers are the most intense of the various branches of the defense industry.

SPECIAL FEATURES OF NUCLEAR WEAPONS CENTERS

The scientific-technical level of the military-industrial complex of the Soviet Union differed sharply from other industries. The technology lag behind the world standard was minimal, if any at all, and many military developments and technologies were at the leading edge of world technological progress. This was achieved with great sacrifice and effort on the part of the entire Soviet people. The level of the military-industrial complex was supported by the redistribution of considerable funds, with the best staff and expensive equipment.

The militarization of Soviet industry was extremely high, its level repeatedly exceeding the official figures. The high social status of defense science and technology was supported by official propaganda, with moral and material incentives. As a result, compared with civilian enterprises, the complex had many advantages in technology, equipment, staff, organization, labor and technological discipline. The striking contrast between military and civilian application was evident in many enterprises. Civilian and defense production shops operated under different conditions at the same plant. Civilian production shops were housed in 18th-19th century buildings with obsolete and worn-out equipment and unfavorable labor conditions. Defense production operated, to the contrary, in the most favorable conditions, with up-to-date, highly automated production, using the most modern technologies.

Conversion provides an opportunity to use the potential of the defense complex for progress in civilian industry, to reimburse the Soviet people in some measure for the expense and hardship they suffered for the sake of strengthening and defensive potential of the state.

From the first days of the creation of the Soviet nuclear industry it was the darling child of the highest leaders of the country. Its requirements were given the highest priority. Since contrary to common opinion, nuclear weapons are a relatively cheap type of armament, even a comparatively small share of military expenditure made it possible to maintain the science of nuclear weapons and the Soviet nuclear military industry at a high level for three decades.

The most important research and development in this field were carried out at two scientific design centers:

- The Research Institute of Experimental Physics (Arzamas-16), and
- The Research Institute of Technical Physics (Chelyabinsk-70).[1]

Just like the two US national laboratories, Livermore and Los Alamos, these institutes were constantly in close collaboration, and competed with each other. Mutual exchange of information, examination of projects and appreciation of working results, and constant scientific and technical competition enabled the Institutes to maintain a high intensity and quality of work.

The unique structure of the Institutes was determined by what was their unique - and practically their only - main task for many years, namely the design of new types of nuclear weapons, analysis of their effects, and the study of nuclear explosion physics. The Institutes have large scientific research units with highly-qualified scientists: theoretical and experimental physicists, mathematicians and chemists, specialists in gas dynamics and material analyzers, designers and technicians. For a long time the Institutes had first pick of the country's best graduates. Prominent scientists working at the Institutes (including Yu.B. Kariton, A.D. Sakharov, Ya.B. Zeldovich, E.I. Zababakhin, N.N. Yanenko et al.) created their own scientific schools. Research workers at the Institutes have built numerous experimental facilities, such as electron accelerators, fast-pulsed nuclear reactors, shock tubes and powerful lasers. The Institutes are equipped with the best in house-produced computing, measuring and recording techniques.

An important feature of the Institutes is that they perform all stages of nuclear weapon development, from fundamental research to full production, including basic and applied research, design, laboratory examination and actual research, preparation for serial production, and copyrights for invention and exploration.

Nuclear weapons centers have business relations with many other enterprises: with developers, researchers and producers of nuclear materials, with rocket and other military equipment designers, and with serial nuclear weapons producing plants. The Institutes use coordinated design documentation standards, a common set of technologies and a common element base with the mass production plants. One of the major conversion goals for the Institutes is to provide their traditional partners with improvements in civilian production, to help them retain jobs, to develop advanced technologies, and to organize production that is necessary for Russia and is commercially profitable.

SEARCHING FOR DIRECTIONS IN CONVERSION

The first problem in conversion is to find a rational direction for converted work. The same problems face other defense enterprises, but perhaps not in such an acute form. Apart from trivial examples (for instance, manufacturing civilian aircraft instead of military air transport), it is difficult to determine how to develop suitable civilian production without radical reconstruction of the enterprises.

The use of the intellectual wealth and material potential of the nuclear weapons centers to accomplish civilian scientific and technological tasks has been discussed periodically since the 1960s. Every time there was a thaw in the international climate there was a renewed wave of discussion. Leaders and research workers at the Institutes understood it was impossible to continue on the nuclear weapons improvement program, and began to search for other scientific and technical directions that would optimally use the specialties and advantages of the Institutes.

At first efforts were made to look for a single task that would suit the Institutes' capabilities. Having a broad understanding of each other's specialist sections, we understood the advantage we had in dealing with complex problems, and we looked for a task that would utilize this asset to the full. After numerous suggestions, we have reached the conclusion that there is no civilian analogy to the weapons program in its scientific capacity and complexity. Hence, there is no scientific and technical issue that could consolidate the whole range of work we do to the same extent as nuclear weapons development.

We have had therefore to choose several directions for our work that would utilize our capabilities best. By the time of the thaw in the international situation and the onset of perestroika a great number of possible directions had been analyzed. Immediately following the declaration of the necessity for deep conversion the search was intensified. Research workers put forward dozens of proposals and a number of new leaders appeared. They have taken over the leadership of some directions in the work. The most hopeful and realistic directions for conversion have been analyzed.

At present it is understood that although the whole of Russian industry needs to radically improved, the greatest requirement for new modern technologies lies in certain fields:

- In the field of power engineering the requirement is to increase the effectiveness and safety of traditional types of power (primarily nuclear power as far as our Institute is concerned) and the development of new non-traditional types.
- In the field of medicine the requirement is the creation of medical equipment with scientific sophistication (for us this means mostly diagnostic medicine and radiation therapy).

- In the field of communication the requirement is the development of new high-capacity communication systems, particularly fiber optics communication.
- In the field of agricultural products the requirement is developing an increased yield capacity and improvements in food preservation technologies.
- In the field of the environment and the monitoring of radiation and chemical pollution, the requirement is to develop methods for the rehabilitation of contaminated territories and suitable environmental technologies.

The main conversion efforts of the Institute are being concentrated in these fields. The Institute did not have enough qualified specialists for all these directions, but fortunately many young and vigorous scientists have appeared. They have a good scientific background, broad-mindedness, and experience in organizing scientific work. They have managed to master the new directions and to do successful work in them.

The conversion work being conducted at our Institute is rather broad in themes, scale, organizational forms and ultimate production types. Here are a few examples:

- research into greater safety in nuclear power engineering;
- investigation into radioactive and chemically contaminated territories;
- research into the possibility of creating solid electrolyte electrochemical generators;
- equipment for developing fiber optics communication lines, including installations for optical cables production, and equipment designed for communication units and consumers' offices;
- optical lens manufacturing, including photo-chromatic lenses;
- creating equipment for crystal growing;
- designing gamma isotope facilities for medical instrument sterilization and agricultural products processing;
- development of diagnostic medical devices using radioactive isotopes;
- using explosives for material synthesis (for instance, synthetic diamonds), material treatment and increasing well production.

All serious suggestions for conversion require basic research. Russian nuclear centers are able to organize research in the fields of nuclear physics, high-pressure physics and computational mathematics. They have vast experience in these types of research and experimental installations, and in mathematical techniques.

PROBLEMS WITH CONVERSION INFORMATION

The second difficulty with conversion is secrecy and the related issue of information. The special status of the work – its paramount importance for state defense and secrecy to the highest degree – created a regime isolated from scientific circles within the country and especially abroad.[2] The effective channels for receiving information, by taking part in international conferences and exhibitions, were absent. Moreover, direct scientific exchange with foreign colleagues was lacking. Exchange with domestic specialists was extremely restricted.

In the 1980s the prohibition on international scientific contacts was ended. Dozens of specialists from our Institute participated in international conferences, informing international scientific circles about the fundamental and applied research work being conducted at the Institute, and receiving data on modern achievements and problems in various scientific fields. However, financial difficulties, lately connected with Russia's considerable currency shortage, hamper international cooperation. Our ability to subscribe to scientific journals have sharply decreased. Thanks solely to the disinterested assistance of the international scientific community we have managed to maintain a minimum level of scientific contact.

For conversion to succeed, it is necessary to organize the collection of scientific and technical, advertising and reference literature, and to extend participation in domestic and international information events through Russian and international scientific organizations. Every conversion project must provide funds for the supply of information. Up-to-date communication facilities, modern information-keeping and processing methods, and the organization of access to international scientific databases are also of great importance.

ORGANIZATION OF CONVERSION WORK IN THE INSTITUTE

Conversion not only changed the orientation of investigations and developments, but necessitated other changes as well:

- staff: the redirection of specialists and engineers to certain new specialities;
- organizational: the creation of new permanent and temporary teams, finding new initiative leaders;
- technical: the development of new devices and scientific and technical equipment, as well as purchasing this equipment;

- economic: the search for new funding resources, and development of methods for stimulating conversion work.

We are able to solve personnel problems within the Institute. As a rule, our specialists are highly-educated and can rapidly acquire knowledge in related areas. The organization of much conversion work, however, needs to be improved. There is a rather peculiar system of scientific guidance in the Institute. The primary role in solving many problems belongs in practice to the theoretical physicists, who are experts in nuclear, plasma and high-pressure physics, and are highly experienced in setting up physical experiments, as well as in mathematical calculation and processing the results. It was the theoreticians who usually united the efforts of experimenters, mathematicians, designers, technicians and test engineers. This unique system is unsuitable for many conversion projects: when a particular project is in progress, it is very often necessary to enlarge the original groups, establish new creative teams, and switch specialists between divisions. So, the third difficulty for conversion is connected with the necessity of changing the system of scientific guidance of projects, as well as with the need for scientific leaders capable of organizing new directions for activities.

The fourth difficulty results from the necessity of reorganization. The scheme of organization of the Institute is based on professional principles rather than on disciplines. It incorporates a system of divisions with experts in related specialties: a division of theoretical physics, a division of experimental physics, a division of computational mathematics, a division of hydrodynamics, etc. Conversion has necessitated changes in this scheme, requiring the setting up of new groups that include various specialists working on one basic problem: a fiber optics center, an explosion technologies center, an environmental center, a radiation medicine center, etc. In many cases this reorganization appeared to be very painful because it involved many people.

Present conditions have necessitated the creation of small-sized enterprises, stock-issuing companies attached to the Institute. The contradictory nature of these enterprises and companies is quite evident: formally they exercise a large degree of independence, but there remains in practice rather close connections with the Institute (as the Institute is one of their founders). These enterprises use the experience and possibilities of the Institute, but very often their financial relations are not quite adequate. As a result, certain groups of fellows, even individual fellows, gain advantages that are sometimes not quite deserved. This may cause social tension. But the strictness of the former administrative and financial structure of the Institute impeded progress in many new directions, and some tension would appear to be the inevitable side-effect of making the promotion of conversion possible.

Many (especially the most important) directions in conversion combined the efforts of our Institute with other Russian institutions and organizations, as well as from other former Soviet republics. The previous administrative methods for achieving this integration (helped by the decisions of the Communist Party and the Soviet government) have expired. Attempts at state regulation of this work by means of State codes have also not been successful. For some of the work we have been able to organize integration by our own free will, in the form of an association of state enterprises, but this is not presently legal in Russia. So we have been forced to re-register as commercial companies, with the participation of both state and private enterprises, as well as private individuals.

FINANCING CONVERSION AND INTERNATIONAL COOPERATION

During the early years of perestroika, when the former Soviet leaders decreed the course of disarmament and conversion, the understanding of the transition from the defense complex to civilian industry was rather simplistic. The opinion was expressed that the major drawback of the Soviet national economy was its extremely high militarization, and once this was removed, civilian branches would begin to develop rapidly. This belief influenced state conversion policy; instead of a carefully studied conversion of the defense industry to civilian goods production, funding was simply cut. Defense industries, especially those with high technologies, began to degrade.

Since 1985 several state conversion programs have been developed in the Soviet Union and then in Russia. Unfortunately, it must be admitted that none were really successful. It should be noted that while conversion promises much for the future, it is very costly at present. But this is often not understood. There are no reliable estimates of what funds should be invested in the main directions of conversion, and these directions are themselves not clearly defined. Sources, and optimal ways, of funding are not well determined.

The tasks set for the first years of conversion did much harm to national economic production; they were oriented towards producing the simplest product, requiring no time and effort in product preparation. This led to a drop in the quality of the products. But serious conversion requires research, which can only be conducted at state expense. This research will, of course, eventually cover its expenses - but not in full and not quickly, which is why it is unrealistic to expect support from both domestic and foreign commercial structures. For the same reason, it is unrealistic to expect private funds for fundamental research work. On the other hand, the state should not give money without return to finance a separate production organization; the financial aid should be only in the form of preferential credit. These two parts of conversion are at present distinguished very clearly. It is much easier to obtain budgetary financing for a commercial project that offers quick profit, than for investigation in fields that might greatly influence Russia's future industrial development.

Since there is no public and legal order to determine priority in conversion projects, the adoption of projects is often based on casual acquaintances, and on the initiative and penetration of the people proposing the project. Rampant bureaucracy, discrepancies in legislation, and the unequal status of state institutions as compared with privatized ones are the main obstacles on the way to their conversion. However, up to now the state has been the main source of finance for conversion, and the main condition for its success is an optimized policy of state support.

In those fields where conversion might be most useful there are such obstacles as the absurd rate of exchange of the ruble, which perverts the ratio of raw product cost to production cost, and, associated with it, the unjustified high personal benefit from foreign contacts, business trips, etc. This leads to a situation in which in such socially significant areas as up-to-date medical equipment and communication techniques, it is preferable to buy foreign equipment with hard currency, a course of action that sometimes (but not in all cases) proves advantageous in the short term, but that creates dependence on the Western manufacturer in the long term, if not forever.

The hopes for external economic aid, which the press diligently supports, inflict great harm on conversion and other modern directions in Russian industry. It was understood too late that the volume of expected (but never received) foreign aid is not enough for Russia, and that the direction of that aid does not always correspond with Russian interests. For some reason there is difficulty in understanding the simple fact that no Western country is interested in creating a Russian industry that would be competitive on the world market, and that we must do this ourselves. We should admit, however, that there were no definite promises from our Western partners; these expectations were created and supported within Russia.

Everything said above does not, of course, eliminate the need for Western credits, so long as they are not accompanied by unacceptable economic or political conditions, nor the necessity of broadening cooperation with foreign industrial companies, and of course the need for wide and open scientific exchange.

Funding of Russia's Nuclear Centers is somewhat better than the funding of most of the other Russian research organizations: the level of Federal funding is higher, as well as the degree of attention paid to us by Federal agencies. Recently, the portion of activities financed by Western sources has also increased. To a large extent, these are contracts financed through the International Scientific & Technological Center (ISTC), which was established as a joint effort between Russia, the European Union, the United States and Japan primarily to support scientists who had been involved in the design and development of weapons of mass destruction. In addition, a series of cooperative agreements has been signed between Russian Nuclear Centers and U.S. nuclear weapons laboratories. Cumulative foreign funding is on the order of 10 per cent of the Center's total budget.

These activities cover many research areas, including environmental monitoring (radiation as well as chemical) and rehabilitation of contaminated areas; industrial safety, mainly nuclear power reactor safety and explosive technology safety; control and accounting of fissile materials; non-proliferation; and various advanced technologies.

The value of these activities for Russian Nuclear Centers is in the possibility of reorienting personnel away from defense programs and improving their level of compensation, but primarily it is in establishing business relationships with Western scientists, faster acquisition of scientific information about the status of research in the rest of the world, and in integration of Russian scientists into the world scientific community.

The problem of how to provide Russia with up-to-date methods of production based on advanced technologies may be solved in different ways. The simplest way is to purchase up-to-date methods of production from the West. But this way costs most in hard currency, it does not utilize conversion possibilities, and it perpetuates the technological lag. It may be justified only in those cases where a certain kind of production cannot be done, or is not expedient to be done in Russia. The second option is to create joint ventures. On the face of it, this is very alluring; with the help of a Western partner one can hope to acquire advanced technologies and, because of the cheap labor force in Russia (which is nevertheless highly-qualified) and of cheap raw materials, production costs would be lower than for Western prototypes. This would present the opportunity to penetrate Western markets. Unfortunately, these alluring perspectives are against the interests of the Western partners. They are not interested in the appearance of competitors. That is why joint ventures as a rule remain at the level of so-called screwdriver technology, i.e., at the level of product assembly from Western kits, on which the meager reserves of Russian hard currency are spent. Western partners are not in a hurry to pass on advanced technological equipment and modern designs. Such joint ventures, however, do not provide Russia with jobs for qualified workers or for specialists; they do not use the intellectual potential of workers in defense enterprises, and cannot be accepted as the main way to conversion.

Of course, joint ventures are quite justified for advancing up-to-date Russian technologies into Western markets. If these new techniques enable consumers to obtain products of high quality, then cooperation with a Western partner will be mutually beneficial. Independent ventures by Russian enterprises into external markets is seldom a success.

The third is the most difficult way, i.e., to set up proper production facilities on a world level, utilizing the features of Russian raw materials, domestic standards and parts. The problem is very complex due to the fact that the quality of the products must correspond with analogous up-to-date Western products, and at least in some aspects ought to excel them. Of course, this way does not eliminate the need in some cases to use foreign raw materials and parts, but the main technologies must be developed independently.

In our opinion, this third way meets, to a great extent, the goals of conversion: the intellectual potential of Russian scientists, designers and technicians is utilized, and their professional level is supported and developed; jobs are created in Russia, not in the West, and the share of skilled labor is supported and increased; and dependence on Western deliveries becomes less vital, and interrelations with Western partners can be built on a more equitable basis. Moreover, we consider that the probability of developing basically new technologies and new kinds of production is essentially higher than when copying fixed technological processes.

CONCLUSION

The problems for conversion are in many aspects connected with the general difficulties which are at present being suffered by Russia: reorganization of the economy, a drop in production, and deficiency in finances. The efficient use of conversion could render essential assistance in the creation of an advanced civil industry in Russia. For this purpose substantial initial investment, the creation of a strong scientific base, the correct choice of directions for conversion, and the use of new organizational forms are necessary. The assistance of the world scientific community will be of great importance in providing scientific information, helping to redirect specialists, and in exchanging conversion experience between different countries.

NOTES

1. During the Institutes' history their names changed more than once. We use the contemporary names introduced in the 1960s.
2. This isolation was eased a little by a well-organized scientific and technical information system, and a great number of scientific journal subscriptions, including international ones.

13

The Problem of Conversion in the U.S. Military Electronics Industry

Jack Ruina

INTRODUCTION

U.S. industry as well as the U.S. government is attempting to convert much of the experience and technical expertise from the military products industry to commercial technology and products. In examining the U.S. military electronic research and development (R&D) establishment, I find that it is, for the most part, not suited for the commercial technology opportunities we foresee. Neither government programs to bolster commercial technology nor government-sponsored non-military projects can help create enough demand for resources of the type and scale used in military projects.

As the U.S. defense budget declines, we can expect that funds applied to military R&D will also be reduced. For several reasons, however, reductions of investment in military technology may not be as large as in other parts of the defense budget. First, spending for R&D helps assure a high level of U.S. military technological capability as a means to compensate for reductions in military force levels - a trade-off of quality for quantity. Another reason is the interest in developing military technologies more appropriate for the expected U.S. military role in the post-Cold War period than those that were focussed on a possible massive confrontation with the Soviet Union. A related reason is a reluctance to surrender the technological capability that has been of fundamental importance to the military establishment when we cannot clearly foresee future military needs. Finally, and not the least of the reasons, is a desire to minimize the severity of economic dislocation in this affluent and, therefore, politically influential sector of our economy.

While we anticipate reductions in military-technology programs, we see major needs for innovative technology in many non-military areas. The United States has pressing problems in such neglected areas as energy, environment, health care, education, transportation - to list a few where government action is called for and where there is a need to inject new technologies on a massive scale. In addition to addressing such large public needs, many see an opportunity to utilize the U.S. military R&D capability, particularly in electronics, to bolster the U.S. position in the commercial market worldwide. Recent developments in microelectronics, optical fiber, video display, and data compression technologies have dramatically decreased the cost and increased the flexibility of computation and communication and have stimulated a tide of new products and services for which the consumer demand has been growing rapidly with no slowdown in sight.

The U.S. market share in the production of many commercial electronic devices has, however, been dismal, despite the fact that many of the discoveries, inventions, and demonstrations of technical feasibility and utility of devices occurred in the United States. For example, all U.S. purchased video cassette recorders, video cameras, and almost all TV receivers and home audio equipment are produced outside of the United States, despite the fact that the United States is the largest consumer of these products. Although it would be near-impossible to recapture market share for these particular products, new technologies, new products, and new services offer opportunities for more and different suppliers in the electronics market.

There is, therefore, a strong will to examine the extent to which the U.S. military electronic technology capability can be redirected and restructured to serve the public sector in areas other than defense and also the extent to which it can bolster the health and productivity of the U.S. commercial electronics industry. Military electronics research and development establishments might, on the surface, seem to be particularly 'convertible,' since electronics has broader areas of application than other major areas of military technology such, as nuclear weapons, rockets, missiles, and stealth aircraft. If, indeed, substantial reorientation and restructuring of the existing military R&D establishment can be effected, more rapid reduction of military programs would become more politically acceptable since the human and economic costs of downsizing would be abated.

U.S. MILITARY ELECTRONICS R&D

The superior quality of U.S. military technology is nowhere more evident than in the area of electronics. Whereas the former Soviet Union's technology may have been almost equal to that of the United States in nuclear weapons and rocketry, in military electronics U.S. superiority is unchallenged. The strength of the military electronics industry lies in such products as radar and optical sensors; ground, air- and space-based communications equipment; data processing and displays for command and control, guidance, and navigation systems; systems for air- and space-based reconnaissance and surveillance; specialized computer hardware and software; electronic devices for jamming and anti-jamming and more.

Military R&D efforts cover a wide class of activities that include basic research, exploratory development, prototype development, the development of one-of-a-kind operational equipment, and - finally - work in support of systems already designed and committed for large-scale production and operational use. Also important is the high level of expertise in large-scale systems engineering that has been developed for the military. Many military electronic systems consist of a large and complex collection of different kinds of hardware and software, often located remotely from each other. The ability to integrate such collections of equipment has become a special and extremely valuable aspect of military technology.

Military electronics R&D is done in industry, in special government and not-for-profit laboratories, and in universities, and it consumes a substantial fraction of the approximately $40 billion expended annually in the United States for military R&D. The major industrial companies involved in electronics, such as Boeing, General Dynamics, General Electric (until 1993), Lockheed Martin, TRW, and Hughes, are also major producers of military equipment across the board - aircraft structures, aircraft engines, rocket motors, spacecraft, missiles, etc. Companies whose military business is limited to R&D in electronics tend to be substantially smaller than these giant corporations. Most military electronics research and development done in industry is focused on *product development*. Some fraction may be in exploratory technology, but hardly any is in basic research.

Essentially all U.S. military electronics R&D is sponsored and paid for by the government. Development of new military products is either sponsored by the government directly or is funded as an allowed overhead cost on other government sponsored work; companies rarely use their own funds for that purpose. At times, however, when a government-sponsored R&D effort is likely to lead directly to a major production program, a company may invest much more in the development work than is specifically allocated for such work by the government, in the hope and expectation that a lucrative production contract will come its way.

Much of the military work in *exploratory electronic technology* is done in laboratories owned and managed by the government and by uniquely American institutions - independent university-managed non-profit laboratories, fully funded by the government and generally doing classified work. Some of these independent laboratories are of outstanding quality and are recognized as major national assets in the production of new technology.

One such laboratory in the electronics area is Lincoln Laboratory, managed by MIT. Since its creation in the early 1950s to develop technologies for air defense, the laboratory has grown and directed its work to many other military applications. However, its technological focus remains in such areas of military electronics as radar and optical sensors, computers, and satellite communications, with a much greater emphasis on basic and applied research than on development of prototype equipment. Lincoln Laboratory's budget is now approximately $350 million per annum and it employs about 1,000 scientists and engineers. Essentially all the work is government-sponsored: 85 per cent for the military, the rest for other government agencies on work that utilizes the specific technical expertise developed for the laboratory's military work. What is special about laboratories such as Lincoln is that they can take on very large, classified, high-technology projects, involving scientists and engineers from different disciplines - projects of a nature and scale that traditional university departments are neither organized nor willing to handle.

Defense-sponsored *basic research* in electronics (as in other technical areas) is concentrated in universities and in almost all cases is unclassified, with research results publishable in the open literature. Generally, the defense-sponsored work done at universities is of such basic nature that it would be hard to identify specific long-term applications that are uniquely military. Nevertheless, the Defense Department is the sponsor of the largest fraction of university research in electronics, just as the National Institute of Health is the major sponsor of basic research in biology and the Department of Energy is the major sponsor of basic university research in high energy physics. There is no fundamental reason why much of the military basic research sponsorship in universities could not be shifted to other, new or existing, government agencies. Such a shift would undoubtedly cause some redirection of research and change the emphasis within projects, but would not necessitate any major personnel or organizational changes within universities.

MILITARY VS. COMMERCIAL ELECTRONICS R&D

There are important differences between the military and commercial sectors of the electronics industry that tend to be overlooked, or seriously underestimated, in the enthusiasm about the potential of converting U.S. military electronic capabilities to the commercial marketplace. U.S. military electronics R&D is, for the most part, done quite separately from the commercial electronics R&D. (This is true not only in electronics, but in other technical areas as well.) Usually entirely different organizations are involved, but even when the same company develops both military and commercial products, the two efforts are in almost all cases physically and administratively distinct and for good reason.

To start with, the R&D technical focus is often in very different domains for the two sectors. For example, the expertise developed for such important military applications as over-the-horizon radar, radar and sonar signal processing, extremely low-frequency communication, high-power lasers, and fire control has very little direct application in commercial fields. Also military equipment is designed to operate in environmental conditions far more severe than those required of commercial gear. It must tolerate exposure to severe nuclear and electromagnetic radiation, mechanical shock, extreme temperatures, etc. As has been reported, even coffee cups and toilet seats designed for military aircraft often have severe (but undoubtedly unnecessary) performance requirements to meet - and their costs reflect those requirements. At any rate, much of the military R&D expertise developed to meet such requirements has little commercial relevance. The differences in the technical focus of the two sectors is demonstrated by the fact that in recent years the amount of technology transfer from military to commercial equipment has been quite low considering the immense size and high-quality of the U.S. military R&D establishment.

Concerns about development and production costs differ greatly in the two sectors of electronics R&D. The whole culture of military acquisition emphasizes optimum performance at whatever cost; the point of diminishing returns pursued by the military is much further up the performance scale than in the commercial world. New technologies offering only small marginal improvements, or with limited application, may nevertheless be pursued by the military even when the development costs are high. The continuous upgrading of U.S. strategic delivery vehicles (missiles and aircraft) and the development of the large variety of nuclear weapons that have been in the U.S. arsenal are examples. Also, military sponsors are far more willing to pursue long-range, technically high-risk projects such as space-based high-power lasers than commercial companies and, in fact, consider it absolutely necessary to sponsor such projects in order to assure the maintenance of a superior level of technological capability.

On the other hand, commercial R&D is concentrated on developments where the end product is well-defined and for which there is high confidence of market success. Long-term research and development is considered too risky and too costly, and has a negative impact on near-term profit and loss calculations. Management performance is judged on short-term payoff, and few managers will risk immediate costs for payoffs that might not be evident for ten years. Consequently, there is little incentive for commercial companies to spend money on basic research and exploratory development. Moreover, in the commercial world, having a lead in an emerging technology does not assure getting the major share of any market that may develop. Hence, it is common to see companies bypass long and costly developments and move into a new product or service area only when the technology is well in hand and there is a ready market for that product or service.

It is the case that some products that take a long time to develop commercially can be immensely profitable. For example, the development of xerography was slow and painful and its ultimate commercial success was hardly clear. But generally, such costly long-term developments may be financial disasters and are avoided by industry. For example, the market for the next generation of supercomputers looked promising in the late 1980s and several organizations started work to produce a commercial product. But unexpected technical difficulties (always to be expected in fact), poorer market conditions, and the emergence of new computer technology (massively parallel machines) has put any commercial promise of supercomputers in great jeopardy. A major player - a company funded by a combination of such giants as IBM, Boeing, Ford and Dupont - has already vacated the market after investing hundreds of millions of dollars in the project. Only such highly capitalized companies as those mentioned could tolerate losses of this magnitude and yet survive.

In short, the methods and style of military electronics R&D are inefficient when the commercial measure of relatively short-term payoff is applied. When a long-term and costly investment in a new technology turns sour for whatever reason, it can mean total disaster to a commercial enterprise, but it would hardly put a military service out of business.

Industry can be much less concerned about R&D costs when they are associated with the production of military electronic equipment. Because there is little financial risk in the development and production of military hardware, profit margins as a percentage of sales are characteristically much lower than would be acceptable for commercial products. Profits as a percentage of private investment may be high, however, since often the government contributes to the initial costs of facilities or at least guarantees to cover these costs if for any reason production contracts are curtailed or terminated.

It is interesting to note here that one of the few commercial enterprises that sponsored a sizeable, first-class program of basic and applied research was the American Telephone and Telegraph (AT&T) in its support of the Bell Telephone Laboratories (BTL). But during the time BTL flourished as a research institution, AT&T was a monopoly, with assured revenues, guaranteed profits and no competition. AT&T was able, therefore, to assume responsibility for technical progress in telecommunications with little concern about competitive advantage. How efficient BTL was as measured by its total cost compared to the flow of new, usable products did not have to be an issue. With the breakup of the telephone company, and the competition for communication services becoming intense, AT&T Bell Laboratories now focus more and more on short-term product development at the cost of long-term research. In other words, as far as R&D is concerned, AT&T no longer has the characteristics of a government enterprise, but has become a competitive commercial enterprise with emphasis on costs and near-term payoff; it has undergone a slow but deliberate conversion in its own way.

Turning to the scientists and engineers in electronics R&D, we find that those involved in the military sector tend to be more highly trained (a larger fraction having PhDs, for example) and more highly paid than their counterparts in non-military work. But, as we noted, much of their expertise relates to aspects of electronics technology that have little relevance to non-military products. Technical personnel are organized and managed to handle major large-scale projects consisting of a wide variety of equipment. The very scale and complexity of the systems tend to be substantially larger than they would be for commercial applications, and management structures have been developed accordingly. Also, in general, there is closer and more continuous contact between the developer and the user throughout the development cycle than one finds for commercial products. With the military as both development sponsor and product consumer, there is much less need to speculate about which technologies warrant investment and which products might eventually succeed in the marketplace than is necessarily the case with commercial products.

Once a product is developed and produced, the military consumer is 'committed' and has a high incentive to buy that product. Turning to a new supplier, or perhaps starting afresh with a new development, would incur additional costs or delays, leaving the military consumer with only the choice between buying a product less satisfactory than expected or not buying at all. As an extreme example, when the U.S. Air Force B-1 turned out to be a much more expensive and less capable aircraft than expected, the only alternative to buying the aircraft was to cancel the program, leaving the Air Force without a new bomber - an action the Air Force naturally resisted. In commercial aviation if the choice were to buy a deficient product whose cost exceeds its benefits by the simple measure of expected profit-to-cost ratio, and hence to lose money, there is no question that the product would not be purchased. But without such a simple measure of utility in the military market, there is strong pressure to focus on the benefits of a new technology with minimum concern about costs.

Finally, the supporting infrastructure in military electronics R&D is much more complex and costly than that in commercial R&D. Documentation and reporting, legal requirements, public accountability, and marketing, as well as consumer presence and involvement in product development, are all more involved and very different in the two sectors.

In conclusion, on close examination of just what constitutes the special quality of the U.S. military electronics R&D establishment, we find it is the people (well-trained and highly paid), facilities (oriented to serve special military needs), funding (which includes substantial support for exploratory development and basic research), a tolerance for failure in achieving goals, and a supporting infrastructure tailored for the military sponsor and the military consumer.

Transfer of this total package to serve the commercial market would not result in a capability prepared to compete effectively with what now exists in the commercial electronics industry. Government subsidy of military R&D organizations to enter the commercial marketplace is likely to be wasteful. 'Pushing' the military sector with subsidies, for example, can only be worthwhile if there is also a strong demand for what it can produce. When, during World War II, there was an extraordinary demand for military equipment, the government subsidized conversion from commercial to military work and was prepared to buy all that was produced. Here 'push' and 'pull' worked together for successful conversion. After the war, the accumulated demand for civilian products was so great and production capacity for such goods so meager that conversion back to the civilian sector was relatively easy. But even then many new products developed by aircraft companies intended for the commercial market (e.g., buses, rail cars, aluminum canoes, etc.) failed in the marketplace.

The obstacles to successful conversion are clearly recognized now by many military technology contractors. To cope with anticipated budget cuts, several companies have consolidated their military technology organizations. Martin Marietta's 1992 purchase of GE Aerospace from General Electric is a notable example. Other military contractors have purchased commercial enterprises to diversify their product lines, but keep them entirely separate. Still others are desperately seeking non-military applications for their military technology. Recognizing the difficulties, few have attempted to convert their organizations on a substantial scale to produce civilian rather than military technology. An interesting case is the one of General Motors (the U.S.'s largest auto maker) purchasing Hughes Aircraft Company (a major military technology company) with the hope and expectation that the advanced technology expertise at Hughes would be applied to GM's commercial products. So far, nothing of substance has emerged from this union, and it is not likely anything will in the near future.

Conversion could work, however, for institutions involved in basic science and technology. Military-sponsored, unclassified research at universities can be shifted with little difficulty and with minimal organizational changes.

Military laboratories devoted to applied research and exploratory development would have greater difficulty. As mentioned earlier, the specific skills and expertise, the high operational costs, and the infrastructure found at labs devoted to military R&D are not what is required for commercial R&D. However, since the U.S. government is now interested in increasing its support of so-called 'generic' or 'pre-commercial' technology, some military electronic labs can make useful contributions. The government has already started a small program to encourage cooperative research between industry and non-profit military research laboratories, with some success in sight. But such programs, although useful, will barely utilize the special and unique capabilities these laboratories have for large-scale, high-technology projects of the type now done for the military. The best prospects these laboratories have to contribute to the commercial sector is to work on so-called dual-use technologies (military and commercial). With its declining budgets, the U.S. Defense Department is trying to use commercial off-the-shelf products to a greater extent. The Defense Department is now taking the next step to develop dual-use technologies that would both help the commercial sector and also provide the benefits to the military from any large-scale commercial production that may result. Examples of such dual-use technologies are flat-panel video displays, high-definition video cameras, and devices for massive data storage.

As we move up the R&D scale to prototype development and full-scale testing, the prospects of converting the full capability of the military R&D establishment are dim. The organizations are not right for this purpose, nor is a large proportion of the engineers and scientists readily adaptable to commercial development projects. Private investment is not likely to be attracted by institutions embarked on conversion, and the government's ability to direct development projects for commercial rather than government use has so far not been particularly successful. One has only to remember the anticipated commercial payoff of such costly government-sponsored developments as the SST, the breeder reactor, laser isotope separation, and the space shuttle to conclude that judgments about commercial potential are best left to commercial enterprises.

If the government should provide indirect support of commercial R&D by such means as tax credits, etc., so that the level of commercial R&D activity in the United States is substantially increased, then undoubtedly some small fraction of scientists and engineers now employed on military projects will shift to commercial projects. But they will have to accept both salary loss and serious culture shock. In general, commercial companies are likely to be interested in younger, lower-salaried scientists and engineers rather than retreaded scientists and engineers from the military R&D sector. Such small-scale diffusion of personnel from the military sector to the commercial sector could hardly be considered conversion.

CONVERSION TO NON-MILITARY GOVERNMENT PROJECTS

There is now a strong desire to redirect some of the military electronic R&D efforts to other government-sponsored programs that involve closely-related technologies – air traffic control, space- and ground-based environmental sensing, navigation systems, and intelligent highway systems, as well as large-scale data processing systems. Work in these areas already uses the military R&D establishment for its purposes and if funding is increased will use more of this establishment. Also the large-scale system engineering expertise developed for military projects has direct application to many proposed new current government projects. It would be relatively easy for organizations doing military R&D to shift to other government sponsors interested in the same technologies.

But there are several realities that must be faced in considering conversion from military to other government programs:

- Military programs are more immune to political interference than are other government programs. 'National security' has been invoked to bypass many of the usual political interests that can affect governmental projects. Non-military programs will have a much harder time resisting at least some 'pork-barrel' projects. These may well utilize some of the technical capability developed for the military, but only for programs of questionable utility.

- Many government-managed systems entail purchase of terminal equipment by the user. For example, use of the space-based global positioning system for navigation of aircraft, ships, etc., requires equipping the aircraft and other mobile platforms with electronic terminal devices. Similarly, any government-managed 'Intelligent Vehicle Highway System' will require electronic gear in automobiles, trucks and buses. The large number of electronic terminals, which would account for the major share of the anticipated costs for these systems, are far more likely to be successfully developed and marketed by a commercial electronic organization than by a 'converted' military organization.

- The totality of new non-military government projects cannot possibly utilize a large fraction of the military R&D capability that would be released with a substantial decline in military budgets. Almost all 'high-tech' non-military government research and development programs in space, energy, and environmental research have much smaller budgets than the military and are also likely to suffer even greater reductions than the military.

We can conclude, therefore, that substantially increased government spending in public, non-military projects may be beneficial to the U.S. electronics industry across-the-board, but would not be a particularly significant factor in converting much of the current military electronic R&D establishment from military to non-military work.

14

Conversion of Space R&D

Nikolai Anfimov and Nikolai Severtsev

INTRODUCTION

Over five years have passed since the former USSR declared a policy for conversion of military technology. Soviet defense industries had achieved a high level of development (as compared to civilian branches of the national economy), and the first phase of conversion was accompanied by euphoria. It was believed that the defense industries would manage to revolutionize the national economy within a few years. So far, however, the results are far from what had been expected.

Let us analyze the problems and difficulties faced by conversion of research and development (R&D), taking rocket and space technology as an example. For more than 45 years the Soviet Union has accumulated great scientific and technical potential in the field of rocket and space hardware. Rocket and space technology is one of only a few key fields in the former USSR that has produced internationally recognized scientific and technological achievements; effective applications in the national economy (communications, meteorology, geology, cartography, etc.); and elevation of the technical level of related science-intensive industries (radioelectronics, computers, power systems, etc.).

We can rank the main areas of rocket and spacecraft R&D applications according to the degree to which the accumulated intellectual potential has been realized:

- Development of commercial and science-oriented rocket and spacecraft. This is the principal direction of cosmonautics development, which has been developed from early times on the 'dual use' principle (communications, navigation, cartography, meteorology, etc.).

- Harnessing rocket and space technology methods and approaches for civilian sectors, such as mining, the oil and gas industries, medicine, agriculture, etc.
- Use of unique experimental facilities for solving commercial and scientific problems.
- Use of the 'brain potential' of the scientists in the rocket and space industry.
- Use of the 'material' potential of the rocket and space industry, such as buildings, equipment and materials.

Thorough analysis of all the enumerated areas is not possible here, so we confine ourselves to some comments and discussion of individual examples for each of the above-stated areas.

DEVELOPMENT OF COMMERCIAL- AND SCIENCE-ORIENTED ROCKET AND SPACE TECHNOLOGIES

The key problem for many areas of civilian development is financial backing. As a rule, interested ministries and agencies are not rich enough to pay for the development and updating of spacecraft to fulfill their own requirements, and returns, as compared with the investments in these systems, may be small and stretched out over the long term. Satellite communications and TV systems are in a somewhat better position because commercial interests finance development. Stock corporations and joint ventures with foreign firms are helping realize various space communications projects.

Virtually the whole national economy is vitally interested in developing satellite communications systems. The role of these systems in the information industry grows steadily. Along with geostationary satellite systems, low-earth orbit satellite systems will eventually permit personal communications from any point on the globe using portable terminals. Future communication systems will be operated by commercial organizations, and cost-effectiveness and profitability will be key criteria for their development and operation. This applies equally to fixed and mobile communications systems, space TV and radio broadcasting systems, and warning and distress signal relay facilities, whose operation may be supported by a network of insurance companies and international bodies.

The Earth remote-sensing satellites are of particular interest for assessing natural resources and monitoring the environment. Here also attempts are being made to attract users able to provide financing. For example, the association 'Soyuzinformsputnik' was established to sell local Earth surface photographs of high resolution. In a public relations gesture, Russian businessmen financed the launch of the spacecraft 'Resource-500' in November, 1992. Its re-entry vehicle delivered the Russian people's greetings to the American people on the occasion of the 500th anniversary of the discovery of America by Christopher Columbus.

Internationally compatible technology is a possible future direction in space technology development. In the distant future we would be able to speak about the creation of a World remote-sensing system, similar to the World Meteorological Organization, which would provide, under UN auspices, for mutual exchange of information in the interests of individual countries and the world community. The same system may become a mechanism for maintaining world stability, verifying compliance with military and environmental agreements.

The role of rocket and space technology in solving ecological problems should be specially noted. First of all, it is used to monitor the state of the environment and industrial effluents, as well as providing information support through mobile communications facilities. Second, the instruments and equipment designed for space applications can be used to solve terrestrial ecological problems. In the more distant future rocket and space hardware could be used for solving such pressing ecological problems as disposing of radioactive waste, preventing near-Earth space pollution with debris and used spacecraft, rehabilitating the ozone layer, and influencing climate change and catastrophic weather phenomena.

Other potential space technology applications lie in the field of materials science, such as the deployment of orbital complexes (space stations) for experiments under microgravity conditions and for industrial manufacturing. In some cases, materials and substances thus produced show substantially improved performance owing to weightlessness, but real manufacturing processes aboard spacecraft are complicated by dynamic microgravity. The study of the behavior of materials under microgravity conditions is the most important area of new technology, both orbital and on the ground. Studies conducted on both manned space stations and unmanned spacecraft can guide technological development on the ground. Unfortunately, the manufacture of materials in space on a commercial basis remains a task for the future. There are many projects proposed for 'space factories' to be launched into orbit by various types of rockets right up to the most powerful rocket, 'Energiya.' There are established associations for realization of such projects, but so far industrial manufacturing of materials in space is uncompetitive with ground technology.

A special case of direct conversion of rockets and space technology is the utilization of decommissioned ballistic missiles. In accord with the treaty signed by the USSR and United States on reduction and limitation of strategic offensive weapons, it is necessary to remove from combat duty and destroy more than 1500 ballistic missiles. The liquidation of so many missiles is an expensive and ecologically dangerous job, as has been proved by the liquidation of medium- and short-range missiles in recent years. At the same time, however, the Strategic Offensive Weapons Treaty permits re-equipping ballistic missiles to deliver peaceful objects into space and high altitudes. The results of preliminary research have shown that almost every missile could be modified to launch spacecraft, probes and other payloads for various scientific and commercial purposes.

There are several projects afoot for modifying the SS-18, SS-19, SS-N-8, and other rockets in order to put payloads of 80-5000 kg into orbit or to deliver payloads to altitudes up to 4000 km. There are some very imaginative proposals for utilizing such rockets. Among them are: experiments for collecting information about the geometric structure of tropical cyclone cores (by means of numerous special re-entry vehicles, launched by one rocket); and rocket complexes for rendering quick assistance to people in distress in remote regions of the globe (rockets delivering emergency supplies to crews of aircraft and ships in distress after accidents and shipwrecks). Since such goals would serve the interest of all the Earth's population, rocket complexes would be created under the auspices of the United Nations with joint financing by concerned countries.

Of course, these applications of rockets dismissed from combat duty would require alteration of the warhead and fairing design, algorithmic updating of the control system, provision of electric links with a spacecraft or probe, and other modifications. Judging from preliminary estimates, the time needed to create rocket and space complexes based on demobilized rockets, will be 2-3 years. This re-equipping of ballistic missiles removed from combat duty provides perhaps the most advantageous and finest example of military R&D conversion. There are, however, some difficulties and obstacles; for example, shorter guaranteed service-time after re-equipping, COCOM-type objections, opposition in Western countries to the use of Russian rockets for launching foreign commercial payloads, and so on.

HARNESSING ROCKET AND SPACE TECHNOLOGY FOR THE CIVILIAN ECONOMY

The basic physical laws and mathematical models used in rocket and space hardware are applicable to other technological fields. This allows the application of the achievements of rocket and spacecraft scientists and specialists to many branches of science and engineering. In the West such applications are called 'spin-off.' It is essential that concerned branches of the Russian civilian economy allocate financial means for utilizing rocket and space R&D experience in solving their technical problems.

Here are some examples:

Open-cut mining of some minerals (diamonds, asbestos, etc.) frequently takes place in quarries of such vast dimensions (a diameter of more than one kilometer, a depth of up to 400 meters and more) that stratification effects occur in the air. The most important of these effects is a temperature inversion, when, following an abrupt weather change, the lower air strata become colder than the upper ones. Natural ventilation (for example, wind) is suppressed due to a sharp drop in turbulence intensity in such a stratified atmosphere. In this case exhaust gases from machinery can fill up the quarry within a few hours, raising nitrogen and carbon oxides concentrations above tolerance limits and making human activity impossible.

Thus the problem of developing an artificial quarry ventilation system becomes urgent. Various techniques, such as aircraft engine jets, thermal sources of continuous (burners) or periodic (explosions) action, have been subjected to a comparative analysis. The problem requires solutions to some hydrodynamic and aerological problems (propagation of dynamic and thermal jets, thermals in a vertically nonuniform medium, and others) and laboratory research. Today a set of methodological techniques has been created for choosing venting facilities that are most effective for specified conditions and working out optimal tactics for their use. For example, tests have revealed that aero engines are most suitable for venting super-deep quarries. Calculations show that a facility fitted with a few engines of a modern airliner (of the TU-154 type) would be sufficient for venting quarries of the above-stated sizes. A project development for of adequate facilities is currently underway.

In other examples, specialists in turbulent flows have designed pneumotexturing jets for reprocessing synthetic filaments into bulk-entangled filaments similar to natural yarns. Studies are currently under way to develop jets capable of producing bulk-entangled filaments at enterprises now in operation. Specialists in spacecraft thermal control systems have developed a design for the baking industry of an advanced oven operating with a special high-temperature liquid in heat transport loops. This technology makes it possible to decrease considerably the operating pressure in heat exchangers and, as a result, to extend equipment life.

The scientific and technological knowledge, methodological approaches and specialists' experience in rocket and spacecraft strength have found their application in a number of other civilian industries:

- In the atomic power complex, elements of atomic reactors demanding a high degree of self-protection and safety have become objects of topical studies for strength, reliability and long-life assurance.
- In the mining industry, operating principles of a hydraulic wedge for breaking large rough blocks have been developed.
- In the oil and gas industry, specialists in materials strength have focused their efforts on assuring the failure-proof operation of main and field pipe-lines. The principal method here was the application of strength-testing principles and element-by-element certification of equipment. Complex tests simulate loading regimes in emergency and corrosive medium effects typical of the natural conditions.
- In transport engineering industry the methods of experimental and theoretical research in space hardware have been applied in projects such as mass production of automobile gas cylinder systems, testing of vibration protection systems for loaded elements of shunting diesel locomotives and rotor snow ploughs, and cushioning truck load-carrying elements and platforms against shocks.

There are also important applications in opto-electronics. At present the Central Research Institute of Machine Building (TSNIIMASH), together with organizations in other industries, has developed a set of high sensitivity opto-electronic equipment, designed for measuring and registering radiation intensity and brightness fields of low-glowing objects, gas and dust formations, and backgrounds.[1] The equipment is capable of operating autonomously (on board aircraft, satellites, air balloons) or as an integrated element of various instrumentation systems. We have elaborated the prospects of using this equipment for the following civilian-oriented applications:

- Global atmosphere sounding in support of ecological projects using satellites, aircraft and balloons.[2] In particular, there are prospects for using UV-radiation equipment for monitoring space pollution such as artificial fragmentations, abnormal ozone content in the atmosphere, distribution of propellants, nitrogen oxides and other substances throughout the Earth's atmosphere.

- Development of a new generation of night-vision devices and devices for observing objects in the UV-spectral range. These are based on wide-angle and telescopic reflector and reflection-lens optics and high-tech opto-electronic devices for receiving and registering coordinate positions.
- Development of high-sensitive sensors (photon counters) registering weak material surface radiation to be used in manufacturing (e.g., growing of crystals, production of filaments, determination of material thermophysical characteristics).

In general the use of advanced scientific-technical developments in the field of high-sensitivity opto-electronics would make it possible to create modern devices for mass production to be used in various industries such as medicine, agriculture and meteorology.

Let us note that the use of the scientific and technical potential of military R&D takes place here in small projects, because the users' means are usually rather limited. The principal problem for big R&D groups is the difficulty of funding large-scale projects that would ensure stable work over the long term. Another problem in this conversion area lies in the legal and economic aspects of the commercial use of results that have been obtained in the course of work done for the Ministry of Defense. Under market conditions military customers would like to share in the profits from such R&D conversion.

USE OF UNIQUE EXPERIMENTAL FACILITIES FOR SOLVING COMMERCIAL AND SCIENTIFIC PROBLEMS

Today one cannot develop adequate mathematical models for all processes, rocket and spacecraft flights. As the rocket and space industry took shape, TSNIIMASH and other organizations created a unique experimental base. It comprises wind tunnels, other gasdynamic facilities, plasmatrones, dynamic and strength test stands, and large vacuum chambers. These facilities simulate or reproduce flight conditions, such as force and heat loads acting on an object, space vacuums, space vehicle thermal conditions and many others. The advantage of exhaustive ground testing is obvious if one takes into account the high cost of flight hardware and the unique character of some space flights.

Maintaining the experimental base and numerous test facilities is expensive. Together with the high sophistication of the test base and the great creative skills of its personnel, this has prompted an exhaustive search for applications in different branches of the civilian economy. Rocket and space hardware operation conditions are so specific and extreme, however, that the test facilities built to simulate them turn out to be unsuited for simulating operational conditions of terrestrial machines, assemblies or devices. For example, it is very difficult to find test objects for wind tunnels where air and gas flows have supersonic velocities and a stagnation temperature of thousands of degrees. Of course, such wind tunnels can generate low-speed flows with moderate temperatures, making it possible to test automobiles, trains and wind power units, but it would be like 'crushing a fly with a steam roller'; that is, this process would be rather wasteful from an economic point of view.

A more appropriate project might be to use aerogasdynamic facilities for research and experimental verification of certain novel technologies. For example, unique piston gasdynamic facilities with multichamber compression developed in TSNIIMASH are designed for aerogasdynamic and thermal testing of launcher and re-entry space vehicle models with simultaneous production of high values of such important parameters as Mach and Reynolds numbers; and stagnation temperature. Of great importance is the fact that, unlike shock tunnels widely used in many countries for the same purposes, the piston gasdynamic facilities ensure a rather long run-time (up to one second), which is sufficient for studying gasdynamic flows with chemical reactions (for example, supersonic combustion in advanced scramjet engines). Piston gasdynamic facilities can heat different gases for a short period of time up to rather high temperatures and pressures (2000–10,000 K, 1000-2000 atm). This process, called 'the controlled gas explosion,' may be widely used in civilian applications.

Problems that can be solved using the piston gasdynamic facilities include: surface hardening of materials; development of mobile piston gasdynamic facilities for nonexplosive removal of mineral self-deposits and asphalt-resin-paraffin sediments from the inner surface of oil tubing strings, and pressing out seized parts of submerged pumps and turbo-drills; and restoring oil wells by destroying oil pools. The same technology may be used for producing fine-dispersion powders of high-temperature inorganic compounds (nitrides, carbides, etc.). This is the so-called 'gas-thermal synthesis' method.

Another use of aerogasdynamic facilities is the production of synthetic diamonds. Fascinating possibilities were discovered using TSNIIMASH's large shock funnel for this purpose. A cylindrical tube of

0.5 m in diameter and 200 m in length was used for accelerating metal disks several kilograms in mass up to speeds of 3-4 km/sec. When such a disk hits carbon-containing targets, conditions for forming the diamond phase are created. At the same time the phase reversal is impossible due to fast cooling. As a result, synthetic diamond crystals of a millimeter in size may be produced. Of course, for industrial realization of these technologies special facilities should be designed and constructed, based on the same principles but, at the same time, providing economies of scale for synthetic diamonds manufacturing.

The strength test base is in a more favorable position. These unique facilities make it possible to test large structures (measuring about 20 m) under conditions of static loads of up to 40,000 tons, and vibration loads of up to 80 tons. It is possible to conduct cryogenic-thermal-static strength tests at high temperatures (up to 800–1200 C) or under cryogenic conditions (down to -196 C). The atomic power and transport engineering, oil, gas and coal mining industries show interest in such tests.

USE OF THE 'BRAIN POTENTIAL' OF THE ROCKET AND SPACE TECHNOLOGY

For several decades the Soviet rocket and space industry has been supported exclusively by the government and has been in the public eye. The prestige of specialists working in the industry was quite high. The salary of defense personnel exceeded that of similar civilian specialists. For these reasons the rocket and space industry attracted the most talented students. A high degree of competition for entry to the institutes and universities training rocket and space technology specialists facilitated selection of the brightest applicants. As a result, the rocket and space industry gathered specialists and scientists possessing a rather high intellectual potential.

As a result of the reduction in government financing over several years, the relative level of salary in the rocket and space industry has dropped; now the salary of defense industry personnel is two-thirds that of civilian industry. This has resulted in specialists and scientists leaving the industry; in particular, specialists of the 'middle' level of the hierarchy possessing great creative potential. Very frequently they are generators of ideas, but they cannot be satisfied under the existing financial conditions. These are the people who often leave their permanent jobs, some in search of higher salaries, others owing to the uncertainty of prospects. At the same time the influx of young specialists has been cut off. While this movement of people can be viewed as a form of conversion, the result has been a decrease of the total creative level of the industry.

USE OF THE 'MATERIAL POTENTIAL'

This implies the use of the physical facilities accumulated in rocket and space industry organizations, e.g., buildings, equipment, and computers. Besides the fact that this type of conversion implies a full local termination of the original activity, it turns out to be rather wasteful. Physical plants were designed, as a rule, for specific technologies, so their use for other purposes requires significant renovations. The released equipment is also specialized and only in rare cases is it suitable for other purposes without updating. The new enterprises being created are untraditional, as a rule, and therefore some difficulties arise from the lack of special experience and professional personnel.

Nevertheless, the rocket and space organizations have had to devote some of their premises to unfamiliar manufacturing operations. For example, at TSNIIMASH the mass production of TV set cases was established in a vacant area. The tendency of this process is toward primitive manufacture, a far cry from so-called 'high technology.'

A CASE STUDY OF CONVERSION OF SPACE R&D

Let us examine the peaceful application of a dedicated rocket-space information system (RSIS) as an example of conversion possibilities. There is a possibility of using existing RSIS information for resolving economic problems.

Remote sensing of Earth from space first became technologically possible in the 1960s. Since then it has evolved into an established and steadily expanding industry with commercial applications in many sectors of the world economy. After the U.S. Landsat program (which was privatized as EOSAT in 1985) came the French-based SPOT. Now Russia has joined the competition for commercial remote sensing business with a range of proven satellites and state-of-the-art product offerings that utilize radar instead of the optical sensing of EOSAT and SPOT.

The newest and most exciting star in the Russian fleet of remote sensing satellites is known as ALMAZ, which in Russian means 'diamond in the rough.' A new era of remote sensing dawned in 1987 with the launch of this 18.5 ton satellite bearing a sophisticated radar remote sensing system. With its capability for 15 m radar resolution under all lighting and weather conditions, using a 10 cm microwave band, the solar-powered sensor aboard ALMAZ represents a quantum leap forward in remote sensing technology.

ALMAZ's prototype, COSMOS-1870, was launched into orbit in July 1987 by the Soviet Proton launch vehicle. The satellite itself, developed and built by the NPO Machinostroyenie manufacturing organization near Moscow, is a universal space platform capable of positioning scientific devices weighing up to 4 tons. COSMOS-1870 was the first Soviet high-resolution radar station capable of operating independently of weather conditions and of daylight. During its two years of operation this experimental prototype satellite collected a large volume of radar images of different natural formations across many nations and oceans of the world.

Now the first commercial version, ALMAZ-1, is in operation. Like its predecessor, this satellite carries a radar apparatus operating in the decimeter waveband with a synthetic aperture and high resolution. The synthetic aperture radar (SAR) allows the mapping of routes 20-240 km long. Unlike the C-1870, the radar information from ALMAZ-1 is transmitted to Earth in digital form by a satellite relay transmitter, which expands the capacity of the radar survey and increases the efficiency with which information is obtained. Ground-based digital synthesis then makes it possible to obtain a radar image recorded on the magnetic tape or a photograph from the radio-hologram. In addition, ALMAZ-1 carries a radiometric scanner (RMS) system that permits sounding of the Earth's surface in the 5 cm, 0.8 cm, and infrared wavebands. When signals (images) are received at the ground station they are processed to provide information in the form of photonegatives or photographs on a scale of 1:150,000. ALMAZ provides assured coverage of almost any region on the Earth's surface at intervals of 1-3 days, depending on the mapping region of interest. Using ALMAZ, radar images have been taken of regions that are virtually inaccessible to photography. These products are now available commercially.

The use of a radar device with decimeter range and horizontal polarization for transmitting and receiving ensures that the radar photographs obtained with ALMAZ have a high information content, permitting observations to be made that are of unique benefit to oceanography, geology, cartography and geophysics, as well as agriculture, forestry, and environmental protection. Thus, the ALMAZ SAR yields information that differs from and can augment the information obtained from the earlier, short-lived U.S. Seasat and the U.S. shuttle-borne SIR-A.

A marketing analysis of this system was carried out and the following data were obtained:

- ecological monitoring of the environment in urban regions - 1.5 billion rubles profit a year;
- aerospace processing of agriculture fields for obtaining bio-information - 4 billion rubles profit a year;
- operational detection of forest fires and forecast of new fires allowing extinguishing of them at the initial stage - 1 billion rubles profit a year.

Partial recovery of expenses from the commercial activity is the most preferable financial arrangement and is typical for commercial use of military rocket-space information systems.

CONCLUSION

In spite of the considerable possibilities of R&D conversion in the field of rocket and space technology, the process is proceeding very slowly and often slips.

Our experience with civilian enterprises and organizations has shown that they are interested, as a rule, only in small-scale projects and, moreover, in projects as short-term as possible that culminate directly in mass production, such as prototypes of technological devices or instruments. Therefore, research teams have to make their research projects small-scale and rearrange their work style to establish numerous small groups of specialists engaged in converting study results into practice.

It is possible to conclude that R&D cannot count on sufficient financial support from the civilian economy. Civilian-only firms pay for work that is in a stage of completion such that a technical device can be used directly by the customer or can be delivered to the manufacturer. As a rule, however, there is no such completion phase without a prior developmental phase. Therefore, the main problem of R&D conversion management is: Where to get money for the first stage of commercial development? Here, it seems, other funds are the most rational financing sources:

- governmental assignments to assist converted enterprises. Such assignments would be given out by the government as compensation for the costs of conversion;

- agency funds for support of concrete conversion projects. Each defense industry branch would finance some concrete projects according to the general branch conversion program;
- enterprise funds formed at the expense of the basic activity;
- international funds.

The last source is still not widely employed in Russia. But of late the situation has been changing. Certain prospects are connected with establishment of the International Scientific & Technological Center (ISTC) in Moscow late in 1992. This center is designed for special financing of conversion projects, proposed by nuclear and rocket specialists from funds allotted by the United States, the European Economic Community and Japan. However, ISTC implementation has been delayed. Only in December 1993 was the center legalized by Russian presidential decree. As a consequence, real financial support for conversion R&D projects did not start until 1994.

Clearly, the most promising avenue of space conversion is the development of commercial and science-oriented spacecraft. Less promising are applications of space technology methods and test facilities to civilian economic problems. Unfortunately, such possibilities exist only for a small portion of extant space technology. The least promising prospect is harnessing the 'brain' and 'material' potentials of the rocket and space industry for entirely unrelated applications. This way leads, in practice, to the loss of decades worth of accumulated rocket and space industry potential.

NOTES

1. This equipment has a spectral range from 0.1 to 0.9 microns and sensitivity up to 0.01 W/steradian (by radiance) and 10^{-9} kW/cm^2 (by brightness).
2. The opto-electronic equipment provides spatial resolution not worse than 1 min and sensitivity per resolution element of 10^{-18} - 10^{-17} W with 300-500,000 image resolution elements.

PART IV

The Way Forward

15

Preventive Arms Control as a Prerequisite for Conversion of Military-Related R&D

Jürgen Altmann, Wolfgang Liebert, Götz Neuneck, and Jürgen Scheffran

INTRODUCTION

Throughout history, military innovation has provided decisive advantages in battle.[1] Particularly in the nuclear age the major powers have pursued superiority by very large efforts devoted to research and development (R&D) for new weapons. In many cases, however, the net outcome was a situation less stable and less safe than before. Even when the opponents began to limit numbers of weapons and carriers, qualitative 'improvement' continued at an undiminished pace.[2] Researchers have pointed out the essential role of military R&D in permanently undermining equilibrium and stability.[3] A comprehensive process of disarmament and confidence-building has thus to include military R&D. Both conversion of R&D facilities to civilian purposes and limits on certain military-related R&D activities seem necessary in such a process.

The process of converting military R&D to civilian purposes can only work well if the whole security environment provides sufficient incentives to disarm and convert. Especially after the Cold War, economic reasons and criteria should be more easily applicable, since the strongest security motives no longer stand against them. Given such an environment, one aim of converting military R&D must be to block future qualitative arms races that would not only divert funds, but could also create new threatening and instable situations.

In order to work towards comprehensive security the existing arms control and arms export control processes need to be examined and extended by preventive measures, supported by technology assessment tools. Disarmament, preventive arms control - especially restrictions on militarily related R&D - and non-proliferation are necessary conditions for an irreversible conversion of military R&D. Conversion and R&D restrictions are complementary paths for preventing dangerous military-technological innovations and proliferation. In this chapter we shall present some ideas along these lines.

THE NECESSITY OF QUALITATIVE AND PREVENTIVE ARMS CONTROL

Traditional arms control has not paid sufficient attention to the issue of qualitative improvements in the arms build-up. In the past the guiding principles were parity and the establishment of upper quantitative limits rather than the aim of (complete) disarmament. Qualitative arms control, on the other hand, would focus on characteristics of types of armaments and would ban complete classes of weapons and carriers.[4]

Preventive arms control can be seen as qualitative arms control applied to the (near or far) future. It would ban qualitative 'improvements' of existing weapons systems or armaments of new kinds that have already been tested, but not introduced, or that may come into being in the future. The concept of preventive arms control aims at cutting off the military-technological innovation process as early as possible. Furthermore, preventive goes beyond qualitative arms control in that it addresses also the danger of the world-wide spread of most destabilizing military technology before the scientific-technological prerequisites for proliferation have been established in more states.

Despite the preponderance of quantitative aspects, some existing bi- and multilateral arms control treaties contain a few commitments that one could call first steps toward qualitative or even preventive arms control through restrictions on the testing, use, manufacturing, production and acquisition of mainly weapons of mass destruction. Some agreements contain explicit restrictions on the development and testing of specific weapon systems and components, others influence R&D and testing indirectly, e.g., a ban on the deployment of certain systems reduces military incentives for R&D in that field (Table 15.1). It should be mentioned that the recent European agreements - the Conventional Armed Forces in Europe (CFE) Treaty of 1990 and the Vienna Document of the Conference on Security and Co-operation in Europe of

Table 15.1
Arms Control Treaties that Include Qualitative Aspects or Have a Direct or Indirect Impact on Research, Development or Testing[a]

Treaty	Year	Obligation Concerning RD&T	Verification Method
Antarctic Treaty	1959	no weapons tests in Antarctica	open access
Limited Test Ban Treaty	1963	no nuclear tests above ground	(NTM)
Outer Space Treaty	1967	no nuclear weapons in orbit no weapons tests on moon and celestial bodies	(NTM), open access on moon and celestial bodies
Latin-American Nuclear Free Zone	1967	no testing and manufacture of nuclear weapons	IAEA safeguards, inspections
Non-Proliferation Treaty	1968	non-nuclear weapons states: no manufacture of nuclear weapons	IAEA safeguards
Biological Weapons Convention	1972	no development of biological weapons	(NTM) (discussions on augmentation underway)
Anti-Ballistic Missile Systems Treaty	1972	limits on land-based ABM systems, no development and testing of sea-, air-, space- or mobile land-based ABM systems	NTM
Threshold Test Ban	1974/90[b]	no underground nuclear test above 150 kt TNT	NTM, inspection
Peaceful Nuclear Explosions Treaty	1976/90[b]	limits and procedures for 'peaceful nuclear explosions'	NTM, inspection
Strategic Arms Limitation Talks (SALT 2)	1979	restraint in development of some new types of strategic offensive arms	NTM
Intermediate-range Nuclear Forces Treaty	1987	complete ban on ground-based missiles between 500 and 5,500 km range	NTM, inspection
Strategic Arms Reductions Treaty (START 1)	1991	restrictions on production and testing of new types of strategic offensive arms	NTM, inspection, exchange telemetry data
Chemical Weapons Convention	1993	no development and production of chemical weapons	inspection, continuous monitoring instruments

[a] The year given refers to signature; some of the most recent treaties are not yet formally in force. Some agreements are bilateral, others are multilateral. If a treaty does not mention verification methods explicitly, this is indicated by parentheses.

[b] Verification Protocols were concluded in 1990.

Abbreviations: ABM: anti-ballistic missile system. IAEA: International Atomic Energy Agency, NTM: national technical means of verification.

1992 - do not contain qualitative, let alone preventive, limits and do not affect R&D. The obligations of the Vienna Document to submit annual deployment plans and to demonstrate new types of major weapon and equipment systems work only after R&D and testing have been completed.

In the present situation after the Cold War, there are specific reasons for introducing or strengthening preventive arms control. One lies in the increased efforts of the highly industrialized nations, above all the United States, to use high technologies for war-fighting purposes.[5] Some developing countries follow dual-use concepts for their own arms development and production, similar to northern countries. With currently more than 50 military conflicts, a third reason stems from the demand of developing countries to procure high-technology weapons.

CONCEPTUAL ISSUES

When thinking about preventive arms control, one has to deal with several conceptual issues. These have to do with the relation between conversion and preventive arms control, with those parts of R&D which - though not declared military - are militarily relevant, and with the role of the military in general, and of weapons innovation in particular, in the future.

Conversion and preventive arms control are complementary. Irreversible conversion of R&D is unthinkable without R&D restrictions, which have to be introduced simultaneously. Preventive arms control would include the whole process of weapon creation, from certain types of research via development to testing and deployment.[6] It would cut new military-technological options at their sources and would thus open the way for conversion of related and obsolete R&D. This would also contribute to creating an international climate of increased security and would reduce mistrust and thus provide incentives for conversion. A debate on adequate preventive arms control would sharpen the view of scientists about which technologies are stabilizing and which are not. The result of such a process - the prohibition of specified weapons - should simultaneously exclude undesirable paths of conversion (e.g., restrictions on space nuclear power should also exclude related R&D for potential civilian applications). The development of new verification methods for old and new arms control regimes provides new tasks for scientists and engineers from the Cold War military establishments, being a form of conversion in itself.

Conversion and preventive arms control meet different difficulties at the various stages of R&D. Conversion seems to be easier at an early (research) stage in the creation process of new weapons, when the path to military or civilian applications is less clearly predetermined. It is just that indeterminate quality, however, that makes preventive arms control more difficult there. The opposite situation pertains at the later stages of development and testing, when the military application dominates: here conversion is more complicated, whereas limits are more easily applied. In certain cases, however, even at the research stage one may be able to find specific criteria for preventing or limiting unwanted military-related work. Likewise a need for conversion exists for the institutions active in the development and testing stages.

Restricting only declared military R&D is not sufficient, since part of the civilian R&D base is militarily relevant and contributes to the military technological innovation process.[7] This holds all the more if dual-use concepts are pursued that organize the support of civilian-labeled R&D projects for simultaneous military development. In addition, including civilian R&D in the limits prevents continuation of banned military projects under a civilian guise. Thus the relevant concept for preventive arms control is 'military-related R&D.' When civilian R&D is included under limits motivated by preventive arms control, however, it should be done with the goal of minimizing impediments to non-military-related work, and legitimate private information should be protected.[8]

As long as intentions, plans or options for new military technology exist, military-related R&D will be pursued and will impede efforts to convert R&D in relevant areas. This could result in diversification instead of conversion, with military R&D programs re-labeled to include 'promising' civilian applications in order to improve the economic competitiveness of such R&D. But behind that is the concept of maintaining the military R&D, base with the possibility of returning to an emphasis on military projects at any time.[9]

More generally, as long as states will find armed forces necessary for the eventuality of war there will be a motive to enable them to fight as effectively as possible and thus to provide them with the most modern equipment. This provides a strong resistance from states and their militaries against comprehensive limits on military-related R&D.[10] Since much of the effectiveness of armed forces in war depends on secrecy about weapons, deployment and strategy, there will be opposition to too far-reaching transparency in military matters in general, and in military-related R&D in particular, which will impede agreement to intrusive confidence-building measures or verification regimes.

Thus, conversion of and limits on military-related R&D can work successfully only within a comprehensive process of disarmament in which the role of the armed forces is continuously being reduced while political and juridical ways of solving conflicts take over. The challenge for arms control and verification research in the field of military-related R&D lies in creatively looking for ways to expand limits, to increase openness and to build confidence, while at the same time conciliating the desire of the militaries and governments to preserve what they feel to be essential capabilities and important secrets. International seminars on military doctrines and adherence to the principle of defense sufficiency may contribute to softening these positions over time.

Finally, there are the questions - at the moment, somewhat hypothetical - of whether in a world gradually renouncing national military force a supranational (UN) force for peace-keeping or peace-enforcing is required, and if so, whether it would need modernization, thus legitimizing a certain amount of military R&D. We feel that this question needs more time and much more study, but it seems that if such R&D were done in an international, agreed setting, it could be done in a limited, circumscribed way and destabilizing consequences could probably be prevented.

STRENGTHENING AND EXTENDING QUALITATIVE AND PREVENTIVE ARMS CONTROL

Including qualitative aspects into existing arms control regimes is necessary on a regional as well as a global level. One approach could be to integrate future arms control issues into regional security systems. Bilateral treaties should be clarified and strengthened where they concern qualitative aspects, the most important example being the ABM Treaty of 1972. In order to avoid a situation where states not formally bound by such treaties transgress their limits and thus undermine compliance by the United States and Russia/CIS, the traditional concept of arms control, created within the framework of the Cold War, should be extended to multilateral approaches.

The end of the Cold War gives unprecedented opportunities to scale back the oversized arsenals of the nuclear superpowers and dampen the proliferation of weapons of mass destruction. Concerning nuclear disarmament, the Nuclear Non-Proliferation Treaty (NPT) should not only be interpreted as stemming horizontal proliferation, but also as clearly aiming at nuclear disarmament and at stopping the dynamics of qualitative improvement of nuclear arsenals, according to its article VI and the preamble.[11] Additional agreements should be envisioned which, if well drawn-up, would directly or indirectly lead to reductions of the relevant R&D:

- a comprehensive nuclear test ban;
- an agreement subjecting all nuclear facilities that are unsafeguarded or unreported so far, in particular those of declared and de-facto nuclear weapon states, to international inspections;
- the halt of the production (and of civilian use) of nuclear-weapon-grade materials, except for insignificant amounts; and
- international storage of all existing weapon-grade material with the goal of finding long-term disposal options making the materials irretrievable for weapon use.

Should in the long run a Nuclear Weapon Convention replace the NPT, the realization of a nuclear-weapon-free world would lead to a marked re-orientation of R&D for nuclear weapons.

Concerning delivery systems, the following steps would have a dampening impact on military-related R&D:

- strengthening of the ABM Treaty and international restrictions on ballistic missile defense systems;
- a treaty banning space weapons (including anti-satellite weapons);
- the inclusion of aircraft in regional arms control regimes;
- global restrictions on the production of ballistic missiles and stringent constraints on missile flight tests;
- expansion of the INF Treaty leading to a world-wide ban of ground-based missiles above 500 km range (with obligations for the existing nuclear-weapons states to reduce their ballistic missile arsenals drastically); and
- a worldwide agreement that includes the potential recipient countries to stop the proliferation of missile technologies and to introduce international verification.

In the chemical weapons field, the Chemical Weapons Convention of 1993 flatly bans development and provides an effective verification regime. The Biological Weapons Convention also bans development, but verification mechanisms still have to be added.[12] Research is not covered in either convention because of difficulties of definition and verification, but creative approaches could lead to effective limits at that stage (for examples in other fields see the following section).

As mentioned, in the area of conventional weapons there are practically no qualitative, let alone preventive, limits in existence. With the Cold War over and with many restrictions on weapons of mass destruction in place, the main thrust of military-technological development will be in the conventional field, and new qualitative agreements and limits will be needed for existing weapons, as well as for R&D for new ones. The following section will give some possible approaches for the latter.

LIMITS ON R&D OF SPECIFIED WEAPON SYSTEMS

Preventive arms control tries to stop new weapons at an early stage. In order to implement such a concept, a few proposals have been brought forward to put limits on militarily relevant R&D.[13] This is a very complex issue and raises difficult conceptual and practical problems. Whereas in most cases it is possible to decide on the military relevance of R&D activities, the difficulty lies in finding separation lines between civilian and military paths of development. Only careful analysis can disclose such distinctions.[14] It may, however, often be impossible, particularly in basic research. This is obvious in many fields, e.g., in materials science, artificial intelligence or optical computing.

There are, however, examples where a distinction between future civilian or military applications can be made at an early stage. This is the case where new weapons principles raise scientific questions. Research work on propagation of high-energy laser or particle beams over long ranges through the atmosphere, for example, is nearly completely motivated by the military goal of creating beam weapons. R&D projects on electromagnetic acceleration of projectiles nearly exclusively have military goals, namely to create electromagnetic guns that would allow higher muzzle velocities. In most areas of explosion, shock wave, and penetration research military considerations dominate.

Another distinction concerns specific large activities that, if banned, would prevent continuation of research in several areas. Underground nuclear explosions are the most prominent example; with a comprehensive test ban, experimental analyses of the ultra-high-temperature and -pressure plasmas produced as well as of the effects of strong radiation are no longer possible.[15] But one has to note that even with a comprehensive nuclear test ban, some work could continue in the context of inertial-confinement fusion (ICF), which can be looked at as very small hydrogen bombs. Thus, in order to stop such research completely, limits would have to be applied here, too.[16] Also the so-called Above-Ground Experiments (AGEX) - laboratory-scale experiments including contained explosions - are intended to serve to maintain R&D capabilities for nuclear weapon improvements.[17]

It would be a useful task for the United Nations to establish scientific study groups to analyze these matters in much more detail. In parallel, national expertise in assessing the military relevance of R&D programs should be established. Proposals have been made to introduce the requirement for a Defence Technology Assessment (DTA) or technology control statements (TCS) systematically into the decision process on military R&D, with international co-operation.[18] Considering that technology today is intimately connected to basic science, it is also necessary to establish a kind of science assessment with the aim to detect, describe and reduce the use of science for militarily relevant purposes.[19] Depending on the results of the assessments, limits on R&D could be applied in certain circumscribed areas, e.g., nuclear tests, high-energy beams in the atmosphere and in space, inertial-confinement fusion research, laser isotope separation, electromagnetic acceleration of projectiles.

To prohibit special new weapon technologies there is a need for well-defined quantitative and qualitative thresholds with procedures in every area of application for separating legal from banned R&D activities. These will have to hold for military and civilian R&D, possibly with different threshold values and different degree of verification.

In working out the thresholds and verification schemes, the following guidelines could be used: the rules and their verification should be simple; ambiguity should be avoided; a safety margin should exist between the limits and the parameters required for weapons; civilian uses should be restricted as little as possible, but continuation of military R&D under civilian guise should be prevented reliably; possible civilian prospects should be weighed against the risks of military diversion of results. Efficiency, cost and intrusiveness of verification methods have to be considered, and suitable compromises between different goals - e.g., concerning protection of sensitive information versus transparency - have to be found.

Applying these guidelines is not easy and requires a thorough analysis for each area of R&D. For the areas mentioned before, we can give a few examples of how it could be done. One could introduce limits on the following: the yield of conventional underground explosions, with rules for notification and provision for observers above some threshold yield; the power-aperture product for laser beams, and on the average power for particle beams radiating in the open; a ban of laser-driven

Inertial Confinement Fusion (ICF) (allowing only experiments using heavy-ion impact); wavelengths, average powers, etc., for lasers used to separate isotopes; the kinetic energy, mass and velocity of projectiles accelerated electromagnetically.[20]

CONFIDENCE-BUILDING MEASURES FOR MILITARY-RELATED R&D

Besides limits and bans on military-related R&D activities there are several options for confidence-building measures. These may be applied either voluntarily or in a legally or politically binding way. In the following, they are listed according to an increasing degree of intrusion or obligation:

- publish military and military-relevant R&D budgets;
- publish military and military-relevant R&D plans, projects and possible deployment schedules;
- publish military R&D institutions, such as laboratories and test ranges (location, size, general task, special projects, personnel);
- conferences on special issues (e.g., weapon programs);
- publish the results of military R&D;
- inspect laboratories and test ranges;
- monitor the careers of scientists and engineers in military R&D;
- mutual visits in specific regions of high mistrust or tension (e.g., to dispel suspicions about secret nuclear programs);
- planning and restraining of military R&D in international coordination among potential opponents; and
- exchange of scientists, safety personnel, planning staff members.

As mentioned above, the more intrusive measures of this list will only become acceptable at later stages of a confidence-building and disarmament process.

VERIFICATION METHODS

In the conversion of military R&D unilateral and multilateral voluntary initiatives play a prominent role. In order to dispel mistrust, these initiatives should, however, be augmented by binding confidence-building measures as well as preventive arms control limiting military-related R&D, both of which would include effective verification. Verification of confidence-building measures would, as in the Vienna Document 1992, consist mainly of visits and on-site inspections. Some of the methods to be discussed below can also be applied. Verification of limits on R&D activities require various specific methods, most of which will be new. The traditional ones, namely national technical means of verification (NTM) and on-site inspection (OSI), will still have important uses.

NTM can be utilized where R&D activities are large, produce strong signals and/or take place in the open. Examples are nuclear explosions (including their preparations), anti-satellite (ASAT) tests, and beam weapon tests in the atmosphere and in space. NTM can be used to trigger challenge inspections. Capable NTM, however, are only available to a very few countries, and thus cooperative verification methods will have to be included in all multilateral limitation agreements.

Cooperative verification is required when activities take place inside and do not produce strong signatures. Of course, the respective rights and obligations have to be written into the agreements and their verification protocols, and compromises between the protection of sensitive information and the transparency needed for effective verification have to be found. There are several possible methods:

- on-site inspections with technical equipment where appropriate in laboratories or at test sites.
- overflights, which can give valuable information on the purpose of installations (structure, security/protection features, power supplies, effluents). They do not suffer from the absentee problem of satellites, can often monitor below the clouds, and are cheaper.
- remote and medium-range sensors, which can gain relevant information from a distance, avoiding intrusion into sensitive areas: radar for monitoring flight tests of aircraft and missiles; laser radar for high-energy laser facilities; seismic and other sensors for underground explosions; seismic and acoustic sensors for artillery test ranges or for characteristics of land or air vehicles.[21]
- inside sensors, which can detect chemical or biological substances or can record personnel activities, etc.[22]
- use of RD&T choke points: sites where solid fuel for ballistic missiles is handled; wind tunnels; accelerators; test ranges, etc.

These methods could be worked out with participation of scientists of the (former?) military-scientific establishments.

ESTABLISHING INTERNATIONAL CONVERSION

Scientists and engineers who have worked in the R&D of weapons have some knowledge about how to destroy these weapons and manage their elimination. Various international projects have been proposed that would offer an opportunity for some of them to convert their know-how and capabilities to remove the heritage of the Cold War:

- build-up of an international agency to monitor and organize the dismantlement and destruction of nuclear weapons and the safe storage of weapons-grade nuclear material;[23]
- 'Vaccines for Peace' to convert national R&D on biological weapons to international R&D for protection and prevention;[24]
- build-up of an International Agency for Ballistic Missile Disarmament (IABMD) to control the global elimination of ballistic missiles (ZBM: Zero Ballistic Missiles).[25]
- as a long-term project, the creation of a World Space Agency to promote international civilian use of rocket and space technology ('Rockets for Peace').[26]

In all these proposals the problem of renewed use for military purposes should not be ignored. The implications of the 'Atoms for Peace' proposal are known: the spread of nuclear technology for seemingly 'peaceful purposes' was used by many countries as a means to build capabilities for nuclear weapons. To avoid this kind of re-conversion stringent controls that can guarantee civilian use need to be established within states. To facilitate this task, it should be considered whether certain civilian technologies with an inherent potential for military use should be placed under international control, at least partly. The concept of the International Atomic Energy Agency (IAEA) - to help in building civilian national nuclear capabilities of all kinds, capabilities that also have military impact, and to control them afterwards by international inspection by the same organization - can be questioned. The pros and cons of this approach, which combines contradictory goals, need to be carefully examined for other fields of R&D.

A World Space Agency, in particular, could make available useful applications of space flight by putting existing technical resources into an international pool to be used by countries without their own national capabilities. An essential precondition, however, would be that all partners of this pool (including the greater powers) put their capabilities under international control to avoid their military use. This would be important to achieve the deal between disarmament and technological cooperation on a non-discriminatory basis that originally was intended by the Nuclear Non-Proliferation Treaty.

The United Nations could help in promoting the conversion of resources from the military R&D sector to useful civilian applications on the national, regional and global level. The 'Dortmund Declaration,' which comprises the results of the first UN Conference on Conversion held in February 1992 in Dortmund, Germany, states that 'the transformation of military related science and technology activities has to be addressed as a central area of conversion policy.'[27] Appropriate science and technology are required to respond to the world's economic, financial, social and environmental needs. In particular, there is a growing need for continued scientific assessment of global environmental hazards. The available 'peace dividends' expected from the disarmament and conversion process could be used to strengthen international scientific programs towards matching these needs. As part of this process, the transformation of military-related R&D, including know-how, people, technologies, equipment and institutions, could be a cornerstone of successfully implementing conversion strategies.

A recent UN study gives a first broad assessment of opportunities, possibilities and limitations of applying military-related resources, including technologies, for environmental protection and disaster relief.[28] Environmental strategies require a comprehensive set of measures, including protective measures for damage prevention and control; restorative measures for repair and rehabilitation; and remedial measures for environmental compliance and development of ecologically benign and energy-efficient technologies.[29] According to the study, advanced research and technology, which are necessary for responding effectively to the environmental challenge, are available in the civilian and the military sector, but are not sufficiently oriented towards environmental goals. Besides R&D on energy, environmental clean-up, and waste management, information and communication technologies (e.g., sensors, computers, networks, models) in the military sector are seen as highly relevant for conversion.[30]

Some of the limitations of this approach have already been pointed out in the UN Study:

> In general, the nature of the civilian sector and its call on resources and technologies are quite different from the demands of the defense sector. Not only are the techniques, technologies, capital, equipment, and human capital and experience different, but so are the appropriate managerial talents, design emphases and end use of civilian products.[31]

The scientific capabilities offered by defense-related institutions should be adequate to the needs in terms of physical-technical characteristics, costs and desirability. For instance, sensors should be similar in frequency, covered area, resolution and availability; communication systems in the number of channels, connectivity, data rate/information flow; computers in storage capacity, speed or compatibility; models in topic, size, adequacy and relevance. In general, technological requirements are less demanding for environmental purposes (and many other civilian sectors) compared with military characteristics, which often may be overspecified, too complex or not cost-effective.[32]

CONCLUSION

Conversion implies change, sometimes reversal, in the trajectory of technological development towards more useful ends. Not all the military R&D complex can be converted to useful civilian purposes. If it is maintained despite serious doubts in its civilian utility, this may be an indicator of continued military interests. Conversion should diminish the utility of certain military options and improve those of non-military ones. A successful conversion of R&D, including preventive arms control, could serve as a model for international cooperative regimes to promote, transfer, and control science and technology in the civilian sector.

NOTES

1. See, e.g., K. Macksey, *Technology in War* (New York: Prentice Hall, 1986).
2. Important qualitative steps are marked by the fission and fusion bombs, ballistic missiles, multiple independently targetable reentry vehicles; see, e.g., B.T. Feld, T. Greenwood, G.W. Rathjens, and S. Weinberg, eds., *Impact of New Technologies on the Arms Race, A Pugwash Monograph* (Cambridge, MA: MIT Press, 1971); F.A. Long and J. Reppy, eds., *The Genesis of New Weapons*, (New York: Pergamon, 1980); and M. Thee, *Military Technology, Military Strategy and the Arms Race* (New York: St. Martin's, 1986). This process has gone on into the present; see e.g.: U. Albrecht and M. Thee, eds., *Military Use of R&D: The Arms Race and Development*, Special Issue, Bulletin of Peace Proposals, vol. 19, no. 3/4, 1988; and W.A. Smit, J. Grin, and L. Voronkov, eds., *Military Technological Innovation and Stability in a Changing World* (Amsterdam: VU Press, 1992).
3. See, e.g., J. Rotblat, ed., *Scientists, the Arms Race and Disarmament* (Paris/London: UNESCO/Taylor & Francis, 1982); Thee, *Military Technology*.

4. Of course, quantitative and qualitative aspects are intimately connected. In order to use quantitative limits, one normally needs qualitative definitions of what constitutes the system to be counted. On the other hand, qualities to be limited generally have to be defined in quantitative terms, e.g., by calibers, payloads or maximum ranges.
5. See, e.g., W.E. Odom, *America's Military Revolution - Strategy and Structure after the Cold War* (Washington DC: American University Press, 1993); National Academy of Sciences, *Strategic Technologies for the Army of the Twenty-First Century* (STAR 21) (Washington DC: National Academy Press, 1992).
6. W. Liebert and G. Neuneck, 'Civil-military Ambivalence of Science and the Problem of Qualitative Arms Control,' in *Controlling the Development and Spread of Military Technology - Lessons From the Past and Challenges for the 1990s,* ed. H.G. Brauch, H.J. van der Graaf, J. Grin, and W. Smit (Amsterdam: VU Press, 1992).
7. See the contribution by W. Smit in this volume.
8. Cf. 'Annex on the Protection of Confidential Information, Convention on the Prohibition of the Development, Production, Stockpiling and Use of Chemical Weapons and on Their Destruction,' in *Geneva: Conference on Disarmament,* CD/1170, Appendix, 1993.
9. See, e.g., J. Alic, L. Branscomb, H. Brooks, A. Carter, and G. Epstein, *Beyond Spinoff - Military and Commercial Technologies in a Changing World* (Cambridge, MA: Harvard Business School, 1992); National Academy of Sciences, *Strategic Technologies*, pp. 28-29. Furthermore, the STAR 21 report states that 'there is a surprisingly wide agreement' among the different U.S. critical technology lists, whether they refer to long-term national security or to the economic prosperity of the United States (pp. 281-82).
10. But note that there can be wider national security motives to agree to R&D limits, e.g., for reasons of stability, non-proliferation, economic savings. See also W. Smit, 'Conditions for Constraining Military R&D,' in *Verification After the Cold War - Broadening the Process*, ed. J. Altmann, T. Stock, and J.-P. Stroot (Amsterdam: VU Press, 1994).
11. For more details see, e.g., W. Liebert, 'How Should We Proceed with the Nuclear Non-Proliferation Treaty?' 44th Pugwash Conference on Science and World Affairs, Towards a War-Free World, Kolymbari, Crete, Greece, 30 June-6 July 1994.
12. See the relevant chapters in Altmann et al., *Verification After the Cold War.*
13. For a few early proposals, see 'Restricting Research and Development - Comments by J. Prawitz,' and 'Summary of Discussion,' in *Impact of New Technologies on the Arms Race*, ed. Feld et al. More recent proposals are discussed in W.A. Smit, 'Controlling Military Technological Innovation - The Role of Verification,' in *Unconventional Approaches to Conventional Arms Control Verification: an Exploratory Assessment*, ed. J. Grin and H. van der Graaf (Amsterdam: VU Press, 1990); W. Smit, 'Steering the Process of Military Technological Innovation,' *Defense Analysis* 7:4 (1991), pp. 401-415; Brauch et al., eds., *Controlling the Development and Spread of Military*

Technology, especially chapters 4-10; and W. Smit, 'Conditions for Constraining Military R&D,' J. Altmann, 'Verifying Limits on Research and Development – Case Studies: Beam Weapons, Electromagnetic Guns,' and W. Liebert, 'Verifying R&D Limitations – Relevance for Preventive Arms Control and Non-Proliferation,' all in *Verification After the Cold War*, ed. Altmann et al.

14. Liebert and Neuneck, 'Civil-military Ambivalence of Science.'
15. See D. Fenstermacher, 'The Effect of Nuclear Test-ban Regimes on Third-Generation-Weapon Innovation,' *Science and Global Security* 1, no. 3-4 (1990), pp. 187-223.
16. See A. Schaper, 'Arms Control at the Stage of Research and Development? - The Case of Inertial Confinement Fusion,' *Science and Global Security* 2, no. 4 (1991), pp. 279-99; and A. Schaper, W. Liebert, W.A. Smit, and B. Elzen, 'Redirecting and Constraining R&D: the Case of Laser Fusion, Laser Isotope Separation, and the Use of Highly Enriched Uranium,' in *Controlling the Development and Spread of Military Technology*, ed. Brauch et al.
17. M. Kalinowski, 'Military and Dual-Use Research Aimed at Circumventing a Ban on Underground Nuclear Testing,' in *Against Proliferation – Towards General Disarmament*, ed. W. Liebert and J. Scheffran, Proceedings, First Conference of the International Network of Engineers and Scientists Against Proliferation (INESAP) (Münster: Agenda, 1994).
18. See M.A. ter Borg and M.J.W. Tulp, *Defence Technology Assessment – Improving Defence Decision Making* (Amsterdam: Nederlandse Organisatie voor Technologisch Aspectenondersoek NOTA, 1987); W. Smit, 'Defence Technology Assessment and the Control of Emerging Technologies,' in *Non-Provocative Defence as a Principle of Arms Reductions,* ed. M. ter Borg and W.A. Smit (Amsterdam: VU Press, 1989); and H. van der Graaf, 'Past Attempts to Control Military R&D and Weapon Innovation,' in *Military Technological Innovation and Stability*, ed. Smit et al.
19. Liebert and Neuneck, 'Civil-military Ambivalence of Science.'
20. See Altmann, 'Verifying Limits on Research and Development'; Schaper, 'Arms Control'; Schaper et al., 'Redirecting and Constraining R&D'; and Liebert, 'Verifying R&D Limitations.'
21. T.H. Braid et al., 'Laser Brightness Verification,' *Science and Global Security* 2:1 (1990), pp. 59-78; P. Richards, 'Verification of a Comprehensive Test Ban Treaty – A Seismological Overview' in *Verification After the Cold War*, ed. Altmann et al.; and Canada, Department of External Affairs and International Trade, 'Non-Seismic Technologies in Support of a Nuclear Test Ban,' (Ottawa, May 1993); J. Altmann, 'Verification Techniques for Heavy Land Vehicles Using Short-Range Sensors,' in *Verification of Arms Reductions – Nuclear, Conventional and Chemical*, ed. J. Altmann and J. Rotblat (Berlin: Springer, 1989); and J. Altmann, 'Intermediate-Range Sensors for Verification and UN Peace-Keeping,' in *Verification After the Cold War*, ed. Altmann et al.

22. The verification annex of the CW Convention mentions continuous monitoring instruments and on-site instruments in several contexts; see 'Verification Annex, Convention.' The report of the verification experts for the BW Convention includes continuous monitoring by on-site instruments among the potential methods; see Å. Bovallius and R. Roffey, 'Evaluation of Efficiency of Verification Methods for the BWC,' in *Verification After the Cold War*, ed. Altmann et al.; and 'Ad Hoc Group of Governmental Experts to Identify and Examine Potential Verification Measures from a Scientific Standpoint,' Report, BWC/CONF.3/VEREX, Geneva, 24 September 1993.
23. See 'Verifying the Dismantlement of Nuclear Warheads,' Federation of American Scientists (Washington, DC, June 1991).
24. See E. Geissler and J. P. Woodall, eds., *Control of Dual-Threat Agents: The Vaccines for Peace Programme* (Stockholm/Oxford: SIPRI/Oxford University Press, 1994).
25. For a discussion of this proposal see 'Revisiting Zero Ballistic Missiles – Reagan's Forgotten Dream,' *F.A.S. Public Interest Report*, May/June 1992.
26. J. Pike, 'New Regimes for Rockets,' in *Against Proliferation.*, ed. Liebert and Scheffran.
27. See: A. Brunn, L. Baehr and H.-J. Karpe, eds., *Conversion: Opportunities for Development and Environment* (Berlin: Springer, 1992), p. xviii.
28. United Nations, Office for Disarmament Affairs, 'Potential Uses of Military-Related Resources for Protection of the Environment,' Study Series 25, 1993.
29. United Nations, 'Potential Uses,' p. 15.
30. On the latter point see J. Scheffran, 'Environmental Applications of Military Information and Communication Technologies,' in *Conversion*, ed. Brunn et al.
31. United Nations, 'Potential Uses,' p. 18.
32. See J. Scheffran, 'Opportunities and Conditions for Conversion of Military Science and Technology,' Paper presented at the Hong Kong Conference on International Cooperation to Promote Conversion From Military to Civilian Industry, 7-10 July 1993, Darmstadt: IANUS Working Paper 8/1993.

16

Ambiguous Civilian R&D: Preventing Destabilizing Military Applications

Wim A. Smit

INTRODUCTION

The relation between civilian and military technology and its related research and development (R&D) has become an issue for a number of reasons. One is the concern that some countries might use civilian-labeled R&D and technology for military purposes, for instance, for manufacturing nuclear, chemical or biological weapons. Less attention has been paid to conventional weapons technology, though concern has emerged in relation to the Missile Technology Control Regime. This regime tries, by export controls, to prevent the spread of missiles and related technology that could be used for the delivery of weapons of mass destruction. A second reason is the need felt by a number of countries to integrate civilian and military technological developments for economic reasons and to keep a strong defense industrial and technological base. This need is particularly felt by technologically advanced military powers like the United States, the United Kingdom and France. In the former Soviet Union and a number of Eastern European countries the conversion of military industry and military R&D to civilian purposes is seen as one of their most pressing problems. This in turn creates an interest in the relation between military and civilian R&D.

When dealing with the subject of civilian and military technology or R&D, one difficulty is knowing what this distinction actually refers to: are civilian and military technology intrinsically different, or is it mainly a difference of funding resources, or of application? We will return to this question at a later stage when we have discussed a number of cases.

Here it suffices to say that the problem of distinction cannot be solved just by definition: no definition will ever be completely satisfactory. Therefore it seems advisable to start with a problem, and subsequently look at the issues and processes that constitute differences and similarities between what is called civilian and military technology or R&D.

This chapter looks into the relation between military and civilian technology and R&D from the perspective of enhancing international security and strengthening an international arms control regime, rather than from a strictly national security or economic perspective. Enhancing international security, moreover, might be a more durable way of enhancing national security than unilaterally strengthening national defense, which might be perceived by other countries as threatening. First a number of cases in which civilian and military technology or R&D are mutually related, will be described briefly. For each of them some interesting characteristic features of this relationship will be indicated, pertaining both to developmental processes and to resulting technologies.

THE CIVILIAN-MILITARY CONNECTION: SOME CASES

Micro-electronics

None of the major innovations in the field of semiconductors originated in research funded by the military. This is true for the transistor, invented at AT&T's Bell Laboratories in 1947, as well as Texas Instrument's discovery of the integrated circuit (IC), and Fairchild Semiconductor's development of the planar process for mass-producing silicon chips, by the late 1950s. But in all three cases, the further development and production were subsidized and stimulated through early purchases by United States government agencies, namely the Department of Defense and NASA. Between 1952 and 1969, government accounted for over 40 per cent of electronic sales in the United States.[1] This situation changed in the 1970s, after the development of MOS technology (the first Metal-Oxide Semi-conductor chip was developed by Mostek in 1969), and Intel's introduction of the microprocessor in 1972, which made possible highly profitable standard products (like calculators), in contrast to the custom-tailored products for defense. The civilian market soon became dominant and took over the role of driver of technological advancement.[2] The turnover rate became greater in civilian applications than in the military sector, and that, together with specific military performance requirements, was the reason why in 1978 the Department of Defense (DOD) initiated its Very High Speed Integrated Circuit

(VHSIC) program. The aim was not only to close the technological gap between the civilian sector and the military , but also to accelerate the most advanced semiconductor technology, to allow industry to perform directed research into very large scale integration and to incorporate these high-speed, high-density chips into current and future military systems. The DOD had the opinion that 'the direction in which the industry is moving in commercial applications is not the same as we need for military application.'[3] The prime contractors chosen for the VHSIC project were the traditional large computer and military companies: Honeywell, Hughes Aircraft, IBM, Raytheon, Rockwell International, Texas Instruments and Westinghouse.[4] Whereas the end products were to be specialized for military applications, the program would at the same time foster key processing technologies that would be more generally applicable, like lithography, computer-aided circuit design tools and low resistance interconnections.[5] The non-specific military part did not amount to more than 20 per cent, however, and only a small portion of funding was devoted to fundamental technologies (which could not be easily disseminated in any case because of national security concerns). At the same time some companies that were not awarded DOD contracts, like General Electric, started their own in-house 'VHSIC' program for introducing this advanced technology into its own products.[6]

Thus in the area of micro-electronics we see that in the early period the technological breakthroughs came from civilian research programs, but that further development and introduction of the technology were stimulated and subsidized by governmental purchases. This situation changed in the 1970s and 1980s when a large civilian and consumer market came into being. DOD's priority remained specific military requirements and performance, rather than making use of what was offered by the civilian market, and it started its own R&D program of VHSIC. By mainly contracting with traditional large defense companies, the actual introduction of advanced products into defense systems (like radar and weapon guidance) was to be secured. The main overlap between the civilian and the military was in the processing and manufacturing technologies, rather than in the end products.

Nuclear Reactors

Present nuclear reactor technology for electricity generation is an offspring of reactor developments in the 1950s and 1960s for military purposes.[7] In the United States reactor development was dominated by military interests, notably in the development of a Navy propulsion reactor. The choice of a light water reactor for civilian power production was a direct result of the navy program. Also the availability of enriched uranium from the U.S. gas-diffusion plants, built originally for the production of highly enriched uranium for the nuclear weapons program, was an incentive to choose this reactor type for civilian applications. An additional factor was that the use of low enriched uranium in exported light water nuclear power plants would give the United States, as the sole supplier of this material, a leverage in ensuring safeguards against diversion of nuclear material. If reactor safety had been of primary importance, however, quite different reactor concepts might have been developed, as has been observed by the former first Chairman of the United States Atomic Energy Commission, David Lilienthal: '[i]n atomic energy several different technologies were in fact proposed and considered in the beginning. And they were soon put aside.' His judgment is that 'not enough time was taken to test reactor ideas and designs that were radically different from the light water design that quickly preempted the market – alternatives that might have provided us with greater safety and certainly far less complexity.'[8]

Similarly, in France in the 1950s, reactor technology development for civilian and military purposes went hand in hand. The development of graphite reactors and of the plutonium reprocessing technology in the French 'civilian' atomic energy program was directly related to the need for plutonium production for a French atomic bomb.[9] Here not only were R&D for civilian and military applications strongly interwoven, but the resulting nuclear facilities were to a great extent 'dual use.' This is still the case for the French fast breeder reactors Phenix and Super-Phenix, which, in addition to electricity generation, produce super high grade plutonium for nuclear weapon use.

From the foregoing, one may conclude that the early 'integration' of civilian and military nuclear technological developments did not result in civilian products optimized for civilian safety considerations. On the other hand, the civilian-military ambiguity in nuclear R&D and the chosen path in nuclear technological developments was often deliberately pursued to keep the option for military use open, as in the case of France. This ambiguity is still a main concern in preventing nuclear weapon proliferation. In particular sensitive technologies like reprocessing and uranium enrichment lower the technological threshold for acquiring nuclear weapons material.[10]

Lasers[11]

Immediately after the construction of the first working laser in 1960 by Ted Maiman at Hughes Aircraft Research Laboratories – a military contractor – the military showed great interest in its further development.[12] As early as 1962, the DOD was reported to be spending $5 million a year on laser research and development, mainly for detection, ranging, guidance and weaponry. Whereas some of DOD's support went to fundamental laser research at the universities, the bulk of military laser R&D was performed in in-house laboratories and through industrial contracts. In view of the small commercial demand for lasers, this funding was crucial to the advancement of laser technology.

Basic research and development were strongly interwoven in the development of laser technology, with many feedbacks from the application oriented development to basic research questions. Because of the early successes of the use of ruby and glass lasers in target illumination and range-finding, the military services became wedded to a technological path that favored solid-state laser types (mainly ruby and glass lasers), neglecting the use of other possible lasing materials. After these early successful applications, much of the research focused on the development of high energy lasers as 'radiation weapons,' in particular as an Anti-Ballistic Missile (ABM) weapon. When the gas laser was invented it was supposed that the energy densities were too small to produce sufficient pulsed power for applications to ABM systems. So the DOD put little money into the further development of gas lasers, concentrating instead on solving the problems of scaling up solid-state lasers. These laser glasses, tailored to military needs, and the associated glass processing technology did not, however, find much commercial application.[13]

Much resistance had to be overcome within this institutional military setting before a 'technological paradigm' shift to gas laser types, in particular the gas dynamic laser could occur. The latter's working principle – adiabatic expansion of a gas – was completely different and rooted in knowledge obtained from the problems of aircraft and reentry vehicle behavior. One of the breakthroughs to much higher powers in gas dynamic lasers occurred at United Aircraft's Pratt & Whitney Division. The available energy from gas dynamic lasers appeared to be orders of magnitude greater than that from solid-state lasers. That the solution for high-power lasers would come from laser types with a relative low-density lasing medium, for which early experiments, moreover, had not looked very promising was not easily accepted within the interlinked technical and funding community that had focused so many years on solid-state lasers.

This brief description of military laser R&D programs shows the close interplay between research and development. It also illustrates the existence of R&D paths bound to a certain technological paradigm, which is sustained by the network of laboratories and funding agencies involved in the program.

Guidance and Navigation Technology

One area where one might have expected an 'integrated' development is in inertial guidance technology (using gyroscopes and accelerometers) for air navigation and missile guidance. No such integrated development has occurred. MacKenzie has shown that the technology has bifurcated (though not absolutely), with each branch developing in a different institutional context.[14] One branch pertains to inertial guidance and navigation for missile guidance; the other is in the domain of civilian and military air navigation. These different contexts have actually resulted in different technological development paths.

One main reason why the military missile environment and the civilian (and military) market of air navigation have been largely separate is their different requirements: high accuracy and ability to operate under extremely demanding circumstances for the military missiles, versus reliability, producibility and economy (with reasonable accuracy) for civilian and military air navigation. In air navigation, moreover, there was much less reluctance to rely also on external reference (star sighting, radio signals) to compensate for build-up errors in the inertial navigation system during long flights. By contrast, for intercontinental ballistic missile guidance, reducing vulnerability through complete independence from the external environment, had great appeal. The resulting divergence in technological paths illustrates the absence of a 'natural trajectory.' The example also shows how the absence of integration of these paths is sustained by the different institutional settings in which diverging requirements are pursued.

LESSONS FROM THE EXAMPLES

We now draw some conclusions on the nature of the relation between civilian and military technologies.

1. At least as important as the invention itself or the creation of a research idea is whether it will be accepted and pursued in a further R&D program. The development of new technologies has become an organized process that needs substantial funding, which may come from governmental agencies, industry or foundations. When a commercial market is lacking, subsidies and demands from the military may enable

an early development and introduction of the technologies, as in the case of micro-electronics and of lasers.

2. The institutional and organizational context may strongly influence the direction of technological developments. The funding agencies and contractors (including the technologists) can be captured in a technological paradigm that determines the course of technological developments, as in the case of the military supported solid-state laser trajectory. Developments in a civilian context may not satisfy military specifications, and thus trigger dedicated military programs, as the VHSIC program.

3. The institutional and organizational context in which development takes place influences the character of the technology, e.g., through requirements for the intended applications. Though modifications are often possible for different applications, the result may not be optimal, as illustrated by the current nuclear power reactors, which originated from military programs and were not designed with safety as a primary design requirement. Some of the high-power laser projects supported by the military were of little use for civilian applications. The influence of the institutional and organizational context is also evident in the development of inertial guidance and navigation technology.

4. Within a technology program, research and development often cannot be separated but are closely interconnected, as illustrated by the case of the military laser development programs. In such frameworks, making a distinction between research and development is highly artificial. So statements that 'development' might be influenced (steered), but not 'research,' are too general. Though much research is not directly related to technology programs or to some specific application, even there it might be useful to look at the possible military impact of large research programs. Therefore, it makes sense to speak of steering of research and development, and not only of development.

5. The rapid innovation process in civilian industry, driven by technological competition, has many resemblances with the technological arms race, and in a number of fields it outpaces military supported developments. Rather than starting new programs, like VHSIC, the military might consider adapting its systems and requirements to these civilian developments, at least at the component level.

6. Overlap between civilian and military technologies is more likely for manufacturing and processing technologies than for products. This is the case for chip production, but in the laser case the processing technologies for high-power glass lasers were of little use for the type of lasers suitable for civilian applications.

7. Each application generally requires a dedicated effort. For the transfer of technology from one context to another, transformations are needed, either of the technology or of the institutional setting to which it is transferred. The degree to which such transformations are necessary will depend on the technology as well as on the various institutional settings. For example, a considerable overlap exists in inertial navigation technology for civilian and military navigation, deviating, in its turn, from inertial missile guidance technology because of the latter's extreme accuracy requirements. The divergence between civilian and military products generally is stronger for end products than for components, though military requirements may result in divergent specifications at the component level as well.

CIVILIAN AND MILITARY R&D REVISITED

Too often technology is considered as a monolithic entity, particularly in discussions on the possibility of directing the R&D process. It is often necessary to distinguish between various technologies. Some technologies, like space technology, or nuclear fusion, for instance, are developed in large centralized programs. Other developments, like micro-electronics, are much more dispersed. It may also be necessary to distinguish between technology as represented in manufacturing processes or in products. For products a hierarchy can be defined, at least analytically, consisting of the system level, the components and the level of 'generic' technologies.[15] In practice the levels may be strongly interlinked, though some development processes may indeed be aimed at designing a system out of components, or at generating generic technologies.

Funding sources, possible applications, institutional setting and intentions, taken separately or together, are relevant to labelling R&D as civilian or military, but they are not always decisive to this issue. Military-funded basic research, for instance, may have no military applications, even if the intention is there, whereas civilian-funded research may lead to inventions with military applications. Indeed, since the development of technology is a heterogeneous process in which many groups and organizations are involved, it is hard to speak of 'the goal' of a research program without reference to a specific participating group or organization. Organizations may participate for different reasons and only be linked to each other through the R&D that serves their respective goals. Military funded research, moreover, may be aimed not at the development of new weapon systems but of verification technology that is needed to sustain constraints on weapon developments.

It has been argued that the distinction between defense and civilian science and technology does not depend upon any inherent characteristics of the science and technology itself, but refers mainly to who commissioned the work, or subsequently appropriated it (e.g., by imposing security classification). Or, phrased differently, the distinction is an institutional one rather than an intrinsic one.[16] Whereas the institutional element is rightly emphasized, the main drawback of this statement is that it tries to draw a sharp division line between technology and its social context. It thus creates an image of technology as a 'thing' that exists independent of its institutional setting. The examples of nuclear reactors, lasers and guidance technology show that the characteristics of technologies and artifacts are tightly bound with the institutional setting in which they were developed. The nature of a technology can be better understood by not isolating it from this 'social' context. This is one of the important insights provided by modern social studies of technology.[17] It is true for the developmental process ('technology in the making') and for 'technology in use.'[18] In a sense the composition and nature of the social network and the technology that it supports mutually influence each other. This is closely related to Winner's observation that 'artifacts do have politics,' that is, technological artifacts may be biased in particular directions by their political environment or by the social order in which they are developed.[19]

These observations are relevant for the question of how civilian and military technology and R&D are distinct, as well as for the issue of steering or directing R&D. So, rather than trying to arrive at sharp definitions - an attempt that is doomed to fail - it is better to start with whatever is considered to be the problem. One problem might be how civilian and military R&D might benefit each other, or how they might be integrated, which is different from the problem of how to prevent the development of destabilizing weapons (rather than preventing all weapons research). The latter implies the question of steering the weapon innovation process and its related R&D. This issue has been dealt with elsewhere.[20] The treatment here will be limited to the question of how to deal with potentially destabilizing R&D with both military and civilian relevance.

MONITORING, ASSESSING AND (RE-)DIRECTING R&D PROGRAMS TO ENHANCE INTERNATIONAL SECURITY

As concerns the military or civilian applications of technology, the situation is seldom black and white. There are only few examples where a technology has either only military applications and no civilian ones, or the other way around. One example of the former is nuclear weapons, but we have also seen that much of the required nuclear R&D, short of nuclear weapon research itself, has an ambiguity as to civilian and military applications, and may lower the technological threshold for switching from civilian to military applications. Another example is stealth technology, which is only of military use; indeed in civilian air traffic, safety concerns would rather require an increased radar signature.

R&D programs are often ambiguous, entailing the possibility of both civilian and military applications. The likelihood (rather than the possibility) of specific applications depends much upon the institutional setting and the social networks in which the programs are carried out, as well as on the technological transformations needed to switch to other applications.

If one wants to guard against the development of destabilizing technologies from civilian or civilian-military ambiguous programs, some institutionalized form of monitoring R&D and assessing potentially destabilizing developments seems necessary. These assessments should not be limited to the 'technology itself,' but should take into account the 'social' context in which it is carried out, as these contexts often influence the course of R&D programs, and co-determine the likelihood of military applications. From such analyses it may, for instance, appear that rearrangements in the institutional setting might be desirable (for instance performing the R&D only in an international civilian R&D program might help prevent diversion towards military applications).

As the situation is seldom black and white, potentially destabilizing R&D requires a trade-off between civilian benefits and the threat to international security. From the latter perspective it might sometimes be preferable to abstain from certain technologies or to choose a technological path with less prospect for military applications. This may require both national constraints and international arrangements for certain R&D areas. Such constraints are contrary to present policies regarding destabilizing technologies, in which the advanced countries permit themselves unlimited access and at the same time try to block the transfer to other countries. One can find examples of this policy in the nuclear area and in the Missile Technology Control Regime. In the longer term such policies are not very effective, as relevant know-how and equipment, even of advanced technologies, tend to become widely available once they have been developed and have been applied. The gas centrifuge technology for uranium enrichment is a case in point. A more non-discriminatory, symmetric approach is, therefore, desirable.

As an example we will discuss two R&D programs that may have severe implications for horizontal and vertical proliferation but have potential civilian applications as well. These are R&D on Inertial Confinement Fusion (ICF) and on Laser Isotope Separation (LIS).[21] For both, trade-offs should be made between the destabilizing proliferation risks and possible civilian benefits. In this framework also the possibility of different, less risky R&D paths should be explored, as well as the desirability of different institutional settings for these R&D activities.

INERTIAL CONFINEMENT FUSION

Inertial Confinement Fusion (ICF) is one of the two roads that are now pursued towards controlled nuclear fusion. In ICF a pellet of a mixture of deuterium and tritium is compressed to extreme high pressures and temperature, using extremely short and powerful laser pulses. These conditions are in principle sufficient to transform the deuterium-tritium mixture into a plasma and to cause thermonuclear fusion. The other road, which was actually initiated before large ICF programs were started, is Magnetic Confinement Fusion (MCF). Here the confinement of the plasma is achieved through strong magnetic fields. So far neither program has achieved an energy output greater than the energy input. Such a positive energy balance is an absolute requirement for civilian applications like electricity generation, but not for military applications of ICF, as we shall see.

MCF research is not classified and in a number of cases it is carried out in international cooperation (e.g., the European JET program). In contrast ICF research in the nuclear weapon states is to a great extent classified because of its military relevance, and for that reason carried out mainly in national contexts. Knowledge of coupling mechanisms between laser energy and the deuterium-tritium mixture and therefore of pellet design, as well as computer codes, are highly classified. Research, however, is now also carried out in non-nuclear weapons states, though on a smaller scale, and by now it is common knowledge that the most effective coupling process is considered to be an 'indirect drive,' in which laser energy is firstly converted into X-ray radiation, which subsequently compresses and heats the fuel. This is similar to the Ulam-Teller principle of igniting thermonuclear weapons; indeed, the ICF thermonuclear explosions are miniature H-bombs on a laboratory scale. ICF can therefore be used for research on and the design of thermonuclear and new ('third generation') nuclear weapons, as well as for testing the effects of nuclear weapons radiation on military equipment.[22] The main ICF programs in the United States are concentrated in the nuclear weapons laboratories, and have primarily such military objectives. ICF is a tool that, to some extent, can replace underground nuclear tests[23] and can maintain the nuclear weapons expertise in case of a comprehensive nuclear test ban. For these reasons it could undermine the effectiveness of a future comprehensive test ban as a measure for halting the 'vertical' proliferation. The technological demands for these military purposes are less severe than for civilian applications, if only because the former do not necessarily require a positive net energy balance.

ICF also poses a risk of horizontal proliferation, because it provides valuable information for designing a thermonuclear weapon. Knowledge rather than the nuclear fuel (deuterium, lithium and tritium) is still a limiting factor for the spread of thermonuclear weapons to threshold nuclear states.[24] This risk was already recognized at the first Review Conference of the Non-Proliferation Treaty in 1975, where the issue of ICF was raised, but was not pursued any further.[25]

Choices for Different R&D Paths

The largest existing laser system for ICF is the huge NOVA laser system at Lawrence Livermore National Laboratories (LLNL) in the United States, consisting of more than 40 laser beams. After more than ten years of experimenting and upscaling of driver energies, the projected required laser energy to achieve substantial fusion has steadily increased. Plans for the successor to NOVA, the 'Athena' laser, has by now been worked out in detail. The costs for this National Ignition Facility (NIF) are estimated by LLNL at $1.07 billion. NOVA and the NIF make use of a neodymium-glass laser. Because of the required cooling time these laser systems are capable of only producing a few shots per day. This is no obstacle for military experiments, but it is prohibitive to civilian applications like electricity generation, which requires 5 to 10 shots per second. The argument, used in public debates, that laser fusion will open up an important source for energy supply therefore merely fulfills a legitimating role. Within ICF an alternative technological path exists, notably the use of heavy ion beams as a driver. The heavy ion driver would allow for the high rate of micro-explosions required for a fusion power plant, but it could, in principle, be used for military R&D on thermonuclear weapons as well. R&D on this technology is still at a very early stage and success is still uncertain both for military and civilian applications. The Neodymium-glass laser route, however, will almost certainly lead to military applications and not to a fusion power plant.[26]

The technological choices thus are quite clear. If a fusion power plant is the primary goal of fusion research and if one wants to avoid much of the vertical and horizontal proliferation risks connected to ICF, the preferred R&D route is that of magnetic confinement fusion (MCF). Part of the 'proliferation damage,' however, has already occurred, as countries like Japan and Germany have followed the example of the United States and have embarked on laser fusion research programs. It is precisely through this type of research that these countries have developed and acquired computer codes relevant for thermonuclear weapon research. Had the United States in the early 1970s not initiated its costly laser fusion programs, other countries very likely would not have pursued this route either: in technology development a 'bandwagon effect' can be observed, in which the technology leaders are followed by others.[27] The United States in embarking on laser fusion research, however, was primarily interested in its potential use for its own nuclear weapon program, overestimating the effectiveness of its classification procedures and underestimating the bandwagon effect.

But even if the ICF route is pursued there would be a more promising path towards civilian applications - using heavy ion beams - than the Neodymium-glass route, which leads straight to military applications and in which the power plant option is excluded. For the possibilities of implementing such a choice one should look at the organizational and institutional setting - the social context - in which the ICF R&D is embedded. This will be discussed in the last section.

LASER ISOTOPE SEPARATION (LIS)

In the laser isotope separation (LIS) process use is made of laser light to separate isotopes, such as uranium-235 from uranium-238, by selective excitation of one of the isotopes, either in an atomic vapor or in a stream of molecules. LIS, which has been under development since the early 1970s, provides a possible alternative to the existing commercial enrichment technologies, that is the gas diffusion and gas centrifuge technology. There are basically two LIS methods potentially suited for use on an industrial scale: Atomic Vapour Laser Isotope Separation (AVLIS) and Molecular Isotope Separation (MLIS).[28] Both methods carry a proliferation risk, in that highly enriched uranium, suitable for nuclear weapons, can be obtained in only two to three enrichment steps. LIS facilities will also be much smaller than gas centrifuge and gas diffusion plants, so that a clandestine facility would be much more difficult to detect. Concerns about these horizontal proliferation risks were already expressed when the R&D was still at an embryonic stage.[29] Since then a 'vertical' proliferation risk has also emerged, in particular for the AVLIS process, which offers good prospects for enrichment of plutonium: reactor grade plutonium can be upgraded by extracting the plutonium-239 isotope desired for nuclear weapons.[30] This was an attractive option for the United States in the 1980s when it was faced with a shortage of plutonium production capacity for its nuclear weapons program. Another application of AVLIS might be the reduction of the content of plutonium-240 in weapons grade plutonium, to make it more suitable for use in sophisticated weapons. This, however, would at the same time pose a problem to those methods of verification of arms control agreements on nuclear weapons that are based on the detection of neutron emissions from these warheads.[31]

Choices for Different R&D Paths

Programs on LIS for uranium enrichment are now underway in several countries, but most vigorously in the United States. In the 1970s, when there were still firm prospects for an expanding nuclear energy market, the U.S. company Jersey Nuclear AVCO Isotopes (JNAI) spent $70 million on developing an AVLIS process for the production of low enriched uranium for use as reactor fuel. At the same time R&D on AVLIS was carried out at LLNL and on the MLIS process at Los Alamos National Laboratories. The AVLIS process seems to be more suitable than MLIS for plutonium isotope separation, which is probably the reason why the U.S. Department of Energy (DoE) chose in the early 1980s to support and further develop only the AVLIS process at LLNL. DoE has funded this program since 1973, spending more than $100 million per year on it since the late 1980s, with the main part of the funding going to the Special Isotope Separation (SIS) project, dedicated to plutonium separation. A demonstration plant is now in use and R&D is continuing, but the construction of a prototype large-scale production plant in Idaho has been cancelled because of the current (over)supply of plutonium from retired nuclear warheads. JNAI terminated its project in 1981 for lack of support by the DoE.

More or less substantial LIS programs are underway in other nuclear weapons states as well. In France an AVLIS program, similar to the LLNL project, is run by the Commissariat à l'Energie Atomique (CEA) at its research center at Saclay. Another AVLIS R&D project is carried out by the French company COGEMA, aiming at a production plant in the late 1990s. The United Kingdom Atomic Energy Authority, together with British Nuclear Fuels Ltd., also has a program for developing AVLIS technology. Both the Soviet Union and China have R&D programs on LIS, but little is known about the status of these programs. Also in non-nuclear weapons states, like Germany, Japan, the Netherlands, Canada, India, Brazil and South Africa, research is going on. The German company URANIT (Uran Isotopentrennungs Gesellschaft mbH), which up to now engaged in the gas centrifuge technology of the enrichment consortium URENCO, has chosen to develop the MLIS process for uranium enrichment. Its Dutch sister company has also embarked on a R&D program on LIS. These companies have a commercial interest in LIS, as current estimates are that enrichment services using LIS cost less than any of the other enrichment technologies.

LIS is much more ambiguous as to civilian and military use than the ICF program. This makes it more difficult to weigh civilian benefits against proliferation risks. On a global scale there is already an oversupply of enrichment services, and so no need of additional enrichment capacity. On the other hand LIS technology might be cheaper, though in relation to the overall costs of running a nuclear fuel plant the advantage might be marginal. Nevertheless, in the competition for contracts with electricity companies price differences might be decisive. Thus it is primarily the *commercial* interests that are to be weighed against proliferation risks. It has been argued that the LIS technology is too advanced to spread easily to less technologically advanced countries.[32] A similar argument was made for the gas centrifuge, but the spread of this technology - for instance, to Pakistan and to Iraq - has shown that once a technology has been developed and is being used, it becomes widely available. If the proliferation issue is taken seriously, it would therefore be better not to develop either of the LIS technologies. This would require both national self-constraint and international arrangements.

But within LIS there is a choice between different technological routes: AVLIS or MLIS. It appears that the main motive for the AVLIS route is its potential for upgrading plutonium for weapon purposes.

AVLIS therefore entails a risk for horizontal proliferation along two routes, not only via its potential for the production of highly enriched uranium, but also of weapons grade plutonium. In the nuclear weapons states AVLIS might be used for the production of highly purified plutonium-239 and the construction of more advanced nuclear weapons, thus adding to the vertical proliferation problem and complicating verification of limitations on nuclear warheads. From the proliferation perspective, MLIS might therefore be preferred to AVLIS. A non-proliferation policy in which some countries are developing the technology and subsequently try to deny it to others is, however, discriminatory and will help, rather than prevent, the spread of the technology, unless a suitable international control regime is developed.

IMPLEMENTING A MONITORING AND CONTROL REGIME ON R&D

If one wants to avoid the development of destabilizing military technologies and weapon systems, emerging either from military or from civilian R&D, a regime for monitoring and assessment of R&D that would function as an 'early warning' system seems necessary. Identification of such 'sensitive' research might then require constraints or a redirection of research paths, as illustrated above for ICF and LIS. Such monitoring regimes should have an international as well as a national component. International monitoring agencies could be associated with particular international institutions, like the Organization for Security and Cooperation in Europe (OSCE), the nuclear non-proliferation regime, and the chemical and biological weapons conventions. These international agencies should be supported by assessments and information from national agencies or organizations. Moreover, as the 'research system' is a rather dispersed activity, it is important that non-governmental organizations, such as professional associations, see it to be their task to engage in such assessments.[33]

The extent to which a redirection of R&D might be realized will depend on the character of the R&D and related technology, in particular its social (or organizational) structure. Some R&D efforts are highly centralized – funded mainly by government and concentrated in a few laboratories, as in the case of ICF. In other cases, like LIS uranium enrichment, private company research projects may exist in addition to national R&D programs. But even here national governments (and parliaments), supported by international agreements, could still influence the R&D strategies of such companies, sometimes by denying financial support (as in the case of the JNAI), but also by making clear, in advance, that no construction of enrichment facilities based on a particular technology would be licensed. This might discourage such companies from engaging in large R&D programs on such technologies.[34] A much more dispersed institutional and organizational structure may exist for R&D in other technological fields like micro-electronics. Whether and to what extent 'external' incentives might influence the direction of R&D, will depend on the specific social structure and type of interests of the groups involved in such programs. Further analysis is required to explore possibilities. It might, of course, turn out that in some situations interference with a specific ('generic' or 'enabling') technology is neither possible nor desirable. In such cases, regulations at a different level (e.g., the system level) might be more effective.

CONCLUSION

The growing interest in integrating civilian and military technology should not close our eyes to the fact that sometimes such an integration might decrease international security - for instance, through an increased risk of weapon proliferation.

This chapter has shown, through a number of historical cases, a variety of possible relations between civilian and military technological developments. It was pointed out that technologies are not intrinsically military or civilian, but that this feature strongly depends on the institutional setting in which technology is developed. Inertial confinement fusion is a case in point. It was, from its inception, developed primarily as a military technology with energy applications only as a secondary goal. Its institutional setting was (and still is) that of national military laboratories. Development within an international collaborative setting, aiming for a civilian energy technology as a primary goal, very likely would have resulted in a different technological path. Both such an international setting and the different technological choices would have made military applications less likely.

Due to the interwovenness of research and development in many technology programs, it makes sense to look into possible military applications as early as possible in what is often called the 'research stage.' In case of undesirable military technological developments, such programs might then still be directed (both institutionally and technologically) away from an integration of civilian and military technology into directions less inimical to international security.

NOTES

1. R. DeGrasse, Jr., 'The Military and Semiconductors,' in *The Militarization of High Technology,* ed. J. Tirman (Cambridge, MA: Ballinger, 1984), pp 77–104. The Minuteman II project subsidized Texas Instrument's early IC production, and the NASA Apollo program played the same role for Fairchild. Between 1950 and 1960 annual factory sales of consumer electronics products showed no growth at all, while sales of electronic products to the military services increased 650 per cent, in constant dollars. P. Forman, 'Behind Quantum Electronics: National Security as a Basis for Physical Research in the United States, 1940–1960,' *Historical Studies in the Physical and Biological Sciences* 18, pt. 1 (1987), pp. 149–229; p. 160.
2. The government's share of semiconductor output fell from 36 per cent of the total in 1969 to about 10 per cent in 1978. See DeGrasse, 'The Military and Semiconductors,' p. 94.
3. Quoted in B.W. Kubbig, 'Determinants of Spin-off in the Context of SDI: the VHSIC Programme,' in *The Relations between Defence and Civil Technologies*, ed. P. Gummett and J. Reppy (Dordrecht: Kluwer Academic Publishers, 1988), p. 245.
4. Kubbig, 'Determinants of Spin-off,' p. 246.
5. W.J. Perry and L. Sumney, 'The Very High Speed Integrated Circuit Program,' in *Review of US Military Research and Development*, ed. K. Tsipis and P. Janeway (Washington, DC: Pergamon-Brasseys, 1984), pp. 33–45.
6. Kubbig, 'Determinants of Spin-off,' p. 253.
7. For a short history, see e.g., R.G. Hewlett and J.M. Holl, *Atoms for Peace and War, 1953–1961* (Berkeley: University of California Press, 1989), chapter 7. See also G. Thompson, 'The Genesis of Nuclear Power,' in *The Militarization of High Technology*, ed. J. Tirman, pp. 63–75.
8. D.E. Lilienthal, *Atomic Energy: A New Start* (New York: Harper and Row, 1980), p. 99.
9. B. Goldschmidt, *Les Rivalités Atomiques* (Paris: Fayard, 1967); W.L. Kohl, *French Nuclear Diplomacy* (Princeton: Princeton University Press, 1971).
10. A.S. Krass, P. Boskma, B. Elzen and W.A. Smit, *Uranium Enrichment and Nuclear Weapon Proliferation,* SIPRI (London: Taylor and Francis, 1983).
11. This section is based mainly on the excellent article of R.W. Seidel, 'From Glow to Flow: A History of Laser Research and Development,' *Historical Studies in the Physical and Biological Sciences* 18, pt. 1 (1987), pp. 111–47.
12. Two years before, a research project for developing a laser had already been supported at the company TRG by the Advanced Research Projects Agency (ARPA, known for funding challenging, advanced research projects with potential military applications). Ironically, a TRG physicist, Gordon Gold, on whose work the project was based, was not allowed to work on 'his' project, because it was classified, and he did not have the necessary security clearance. See B. Bertolotti, *Masers and Lasers; An Historical Approach* (Bristol: Adam Hilger Ltd., 1983), p. 112.

13. When, for instance, American Optical Company (where the Nd-glass laser was developed) was acquired by a larger firm in the late 1960s, the new owner canceled the laser glass program, because it lacked commercial interest (Seidel, 'From Glow to Flow,' p. 126).
14. D. MacKenzie, *Inventing Accuracy: A Historical Sociology of Nuclear Missile Guidance* (Cambridge, MA: MIT Press, 1990).
15. The levels are not unambiguous: a component may be considered as a system in itself, consisting in its turn of (sub-)components. See W. Walker, M. Graham and B. Harbor, 'From Components to Integrated Systems: Technological Diversity and Integration between Military and Civilian Sectors,' in *The Relations between Defence and Civil Technologies*, ed. Gummett and Reppy, pp. 17–37.
16. P. Gummett, ed., *Future Relations between Defence and Civil Science and Technology*. A report prepared for the Parliamentary Office of Science and Technology, prepared by the ESRC\SPSG Defence Science and Technology Policy team. SPSG Review Paper No. 2 (London, 1991), p. 27.
17. T.J. Pinch and W.E. Bijker, 'The Social Construction of Facts and Artifacts: Or how the Sociology of Science and the Sociology of Technology Might Benefit Each Other,' in W.E. Bijker, T.P. Hughes and T.J. Pinch, *The Social Construction of Technological Systems: New Directions in the Sociology and History of Technology* (Cambridge, MA: MIT Press, 1987), pp. 17–50.
18. J.M. Staudenmaier, *Technology's Storytellers: Reweaving the Human Fabric* (Cambridge, MA: MIT Press, 1989).
19. L. Winner, *The Whale and the Reactor: A Search for Limits in an Age of High Technology* (Chicago: University of Chicago Press, 1986), pp. 19–39.
20. W.A. Smit, 'Steering the Process of Military Technological Innovation,' *Defense Analysis* 7, no. 4 (1991), pp. 401–15; W.A. Smit, J. Grin and L. Voronkov, eds., *Military Technological Innovation and Stability in a Changing World* (Amsterdam/Concord, MA: VU University Press/Paul & Co., 1992).
21. The examples to a great extent draw upon A. Schaper, W. Liebert, W.A. Smit and B. Elzen, 'New Technological Developments and the Non-proliferation Regime. Redirecting and Constraining R&D: The Case of Laser Fusion, Laser Isotope Separation, and the Use of Highly Enriched Uranium,' in *Controlling the Development and Spread of Military Technology: Lessons from the Past and Challenges for the 1990s*, ed. H.G. Brauch, H.J. van der Graaf, J. Grin and W.A. Smit (Amsterdam/Concord, MA: VU University Press/Paul & Co., 1992), pp. 121–38.
22. For an extensive assessment of the military and civilian applications, see A. Schaper, 'Arms Control at the Stage of R&D – The Case of Inertial Confinement Fusion,' *Science & Global Security* 2, no. 4 (1991). For an earlier assessment see W.A. Smit and P. Boskma, 'Laser Fusion,' *Bulletin of Atomic Scientists* 36 (December 1980), pp. 34–38.
23. Whereas ICF research can provide valuable physics insights for developing thermonuclear and 'third generation' nuclear weapons, there is still a gap of a factor equal to one hundred between the explosive yield of ICF experiments

(about one quarter of a ton) and the smallest fission driven nuclear device (with a yield of about 25 tons), causing scaling problems. See R.E. Kidder, chapter 11, this volume.

24. One should realize, however, that for the manufacture of a thermonuclear weapon highly enriched uranium or plutonium are still needed for the required 'fission trigger.'
25. Smit and Boskma, 'Laser Fusion.'
26. See the assessments referred to in note 22. This was also concluded by the Galvin task force (February 1995) on the future of the U.S. nuclear weapons laboratories; see Tom Zamora Collina, 'Livermore on the Defensive,' *Bulletin of the Atomic Scientists* 51 (May/June 1995), pp. 42–45.
27. The 'risks' of heavy investments in such technology developments are also lower for the follower countries in cases where leader countries have already demonstrated some successes. J.H. Fujimura, 'The Molecular Biological Bandwagon in Cancer Research: Where Social Worlds Meet,' *Social Problems* 35 (1988), pp. 261–83.
28. Krass, et al., *Uranium Enrichment and Nuclear Weapon Proliferation*.
29. B.M. Casper, 'Laser Enrichment: A New Path to Proliferation?' *Bulletin of Atomic Scientists* 33 (January 1977), pp. 28–41; Allan S. Krass, 'Laser Enrichment of Uranium: The Proliferation Connection,' *Science* 196 (13 May 1977), pp. 721–31.
30. A. Krass, 'Laser Enrichment of Plutonium,' in *Sipri Yearbook 1982, World Armaments and Disarmament* (London: Taylor and Francis Ltd., 1982), pp. 281–89.
31. S. Fetter, et al., 'Detecting Nuclear Warheads,' *Science and Global Security* 1 (1990), pp. 225–302.
32. For example, '. . . one or more of the AIS [Advanced Isotope Separation] process may prove to be inherently more proliferation resistant than centrifuges, primarily because of the greater technological sophistication required to construct, operate or modify an AIS facility. This may make AIS technologies available to far fewer countries than centrifuge technology'; U.S. Arms Control and Disarmament Agency (ACDA), 'Advanced Isotope Separation and Centrifuge Enrichment,' *Fiscal Year 1981 Arms Control Impact Statements* (Washington, DC: U.S. Government Printing Office, 1980), p. 491.
33. A good example of how this might be done is provided by the assessments studies of the American Physical Society on issues like Nuclear Safety, Nuclear Waste, and lasers for Ballistic Missile Defense.
34. Several examples exist of the influence of external pressures on R&D policies of private companies. For the shift in the R&D trajectory on 'biological' pesticides by the company Monsanto, see J. Jelsma, 'Genetisch Veranderde Organismen in het Milieu: Het Ontwerpen van Veilige Technologie als Interactief Proces,' in C. Daey Owens, et al., *Constructief Technologisch Aspectenonderzoek* (NOTA publication. Den Haag: Staatsdrukkerij, 1987), p. 74.

Index